CLIMATE CHANGE AND ITS CAUSES, EFFECTS AND PREDICTION

GREENHOUSE GAS EMISSIONS

DRIVERS OF INTENSITY AND COUNTRY VARIANCES

Climate Change and its Causes, Effects and Prediction

Additional books in this series can be found on Nova's website under the Series tab.

Additional E-books in this series can be found on Nova's website under the E-books tab.

Environmental Remediation Technologies, Regulations and Safety

Additional books in this series can be found on Nova's website under the Series tab.

Additional E-books in this series can be found on Nova's website under the E-books tab.

CLIMATE CHANGE AND ITS CAUSES, EFFECTS AND PREDICTION

GREENHOUSE GAS EMISSIONS

DRIVERS OF INTENSITY AND COUNTRY VARIANCES

MICHAEL KNOX
AND
JONATHAN D. MARK
EDITORS

Nova Science Publishers, Inc.
New York

For permission to use material from this book please contact us:
Telephone 631-231-7269; Fax 631-231-8175
Web Site: http://www.novapublishers.com

Library of Congress Cataloging-in-Publication Data

ISBN 978-1-62081-287-7

Published by Nova Science Publishers, Inc. † New York

CONTENTS

PREFACE

The interaction of three variables underlie debates on the issue of climate change and what responses might be justified: the magnitude and rates of change of (1)population, (2)incomes, and (3)intensity of greenhouse gas emissions relative to economic activities. This book examines the interrelationships of the variables to explore their implications for policies that address climate change. Both internationally and domestically, initiatives are underway both to better understand climate change and to take steps to slow, stop, and reverse the overall growth in greenhouse gas emissions, the most important of which is carbon dioxide (CO2), emitted by the combustion of fossil fuels.

Chapter 1 - In the context of climate change and possible responses to the risk associated with it, three variables strongly influence the levels and growth of greenhouse gas (GHG) emissions: population, income (measured as per capita gross domestic product [GDP]), and intensity of emissions (measured as tons of greenhouse gas emissions per million dollars of GDP).

$$(Population) \times (per\ capita\ GDP) \times (Intensity_{ghg}) = Emissions_{ghg}$$

This is the relationship for a given point in time; over time, any effort to change emissions alters the exponential rates of change of these variables. This means that the rates of change of the three left-hand variables, measured in percentage of annual change, sum to the rate of change of the right-hand variable, emissions.

For most countries, and for the world as a whole, population and per capita GDP are rising faster than intensity is declining, so emissions are rising. Globally, for the variables above over the period 1990-2005, *the rates of change* (Δ) in annual percent sum as follows (numbers do not add precisely because of rounding):

$$Population\ \Delta + per\ capita\ GDP\ \Delta + Intensity_{ghg}\ \Delta = Emissions_{ghg}\ \Delta$$
$$(+1.4) + (+1.7) + (-1.6) = (+1.6)$$

As can be seen, global emissions have been rising at a rate of about 1.6% per year, driven by the growth of population and of economic activity.

Within this generalization, countries vary widely. (Unless otherwise noted, comments about countries refer to the top-20 emitters as of 2005, who accounted for about 75% of world emissions that year.) Between 1990 and 2005, in some countries, including Brazil, Mexico, Indonesia, and South Africa, population growth alone exceeded the decline in intensity. For most countries, and for the world as a whole, per capita GDP growth exceeded the intensity improvement each achieved. Countries for whom intensity improvements were greater than their per capita GDP increases included Germany, the United Kingdom, the United States, France, and South Africa. And both the Russian Federation and the Ukraine, following their economic contractions in the 1990s, posted negative numbers for population, per capita income, intensity, and GHG emissions between 1990 and 2005. Besides the Russian Federation and the Ukraine, only the United Kingdom and Germany reduced their GHG emissions for the period (Germany being helped by reductions in the former East Germany).

Stabilizing greenhouse gas emissions would mean the rate of change equals zero. Globally, with a population growth rate of 1.4% per year and an income growth rate of 1.7% per year, intensity would have to decline at a rate of -3.1% per year to hold emissions at the level of the year that rate of decline went into effect. Within the United States, at the 1990-2005 population growth rate of 1.1% per year and income growth rate of 1.8% per year, intensity would have had to decline at a rate of -2.9% per year to hold emissions level; however, U.S. intensity declined at a rate of -1.9%, leaving emissions to grow at 1.0% per year.

Looking to the future, under auspices of the Copenhagen Accord, the United States has submitted a target of reducing emissions from the 2005 level by 17% in 2020. This would require the United States to reduce the intensity of its emissions by some -4.6% per year during the 2010-2020 decade. This implies that the rate of intensity decline needs to better than double.

Chapter 2 - Nations that are parties to the United Nations Framework Convention on Climate Change periodically submit inventories estimating their greenhouse gas emissions. The Convention Secretariat runs a review process to evaluate inventories from 41 "Annex I" nations, which are mostly economically developed nations. The 153 "non-Annex I" nations are generally less economically developed and have less stringent inventory reporting guidelines. The Department of State (State) represents the United States in international climate change negotiations. GAO was asked to report on (1) what is known about the comparability and quality of inventories and barriers, if any, to improvement; (2) what is known about the strengths and limits of the inventory review process; and (3) views of experts on implications for current and future international agreements to reduce emissions. GAO analyzed inventory reviews and inventories from the seven highest-emitting Annex I nations and seven of the highest emitting non-Annex I nations. GAO also selected and interviewed experts.

Chapter 3 - Using the World Resources Institute (WRI) database on greenhouse gas emissions and related data, this chapter examines two issues. The first issue is the separate treatment of developed and developing nations under the United Nations Framework Convention on Climate Change (UNFCCC), the Kyoto Protocol, and the Copenhagen Accord. This distinction has been a pivotal issue affecting U.S. climate change policy. The second issue is the difficulty of addressing climate change through limiting greenhouse gas emissions to a specified percentage of baseline emissions (typically 1990). The data permit examination of alternative approaches, such as focusing on per capita emissions or the greenhouse gas emission intensity (measured as emissions per unit of economic activity). Key findings include:

A few countries account for most greenhouse gas emissions: in 2005, China led by emitting 19% of the world total, followed closely by the United States with 18%; no other country reached 6%; the top eight emitters (those emitting 2% or more of total emissions) accounted for 58% of the 185 nations' emissions.

Land-use effects (e.g., deforestation) on emissions are negligible for most nations, but they cause emissions to rise sharply for certain developing nations, most notably Brazil and Indonesia.

While countries whose economies are dominated by oil- and gas-production have the highest per capita greenhouse gas emissions, in general developed nations rank high in per capita emissions (in 2005, Australia, the United States, and Canada ranked 6, 9, and 10, respectively, in the world), while developing nations tend to rank low (China, Brazil, Indonesia, and India ranked 81, 84, 117, and 148, respectively).

The greenhouse intensity of the economy—the metric by which the George W. Bush administration addressed climate change, and by which China has proposed to set its objectives under the Copenhagen Accord—varies substantially among developed countries (in 2005, not accounting for land use, Ukraine emitted 512 tons/million international $GDP, while France emitted 80 tons/million $GDP, with the United States at 153 tons/million $GDP; developing nations range from the 136 (Mexico) to 372 (China).

The time frame adopted for defining the climate change issue and for taking actions to address greenhouse gas emissions has differential impacts on individual nations, as a result of individual resource endowments (e.g., coal versus natural gas and hydropower) and stage of economic development (e.g., conversion of forest land to agriculture occurring before or after the baseline).

Differentiating responsibilities between developed and developing nations—as the UNFCCC does—has failed to engage some of the largest emitters effectively. Moreover, many developed countries have not achieved stabilization of their emissions despite the UNFCCC. Given the wide range of situations illustrated by the data, a flexible strategy that allows each country to play to its strengths may be necessary if diverse countries like the United States and China are ever to reach agreement. The difficulty in finding a common strategy was evidenced by the outcome of the Copenhagen meeting, which set a climate change objective of holding global warming to less than 2 degrees C but then left up to each country the choice of how to address emissions.

In: Greenhouse Gas Emissions
Editors: M. Knox and J. D. Mark

ISBN 978-1-62081-287-7

Chapter 1

GREENHOUSE GAS EMISSION DRIVERS: POPULATION, ECONOMIC DEVELOPMENT AND GROWTH, AND ENERGY USE*

John Blodgett and Larry Parker

SUMMARY

In the context of climate change and possible responses to the risk associated with it, three variables strongly influence the levels and growth of greenhouse gas (GHG) emissions: population, income (measured as per capita gross domestic product [GDP]), and intensity of emissions (measured as tons of greenhouse gas emissions per million dollars of GDP).

$$(Population) \times (per\ capita\ GDP) \times (Intensity_{ghg}) = Emissions_{ghg}$$

This is the relationship for a given point in time; over time, any effort to change emissions alters the exponential rates of change of these variables. This means that the rates of change of the three left-hand variables, measured in percentage of annual change, sum to the rate of change of the right-hand variable, emissions.

* This is an edited, reformatted and augmented version of the Congressional Research Service Publication, CRS Report for Congress RL33970, dated March 5, 2010.

For most countries, and for the world as a whole, population and per capita GDP are rising faster than intensity is declining, so emissions are rising. Globally, for the variables above over the period 1990-2005, *the rates of change* (Δ) in annual percent sum as follows (numbers do not add precisely because of rounding):

$$Population\ \Delta + per\ capita\ GDP\ \Delta + Intensity_{ghg}\ \Delta = Emissions_{ghg}\ \Delta$$
$$(+1.4) + (+1.7) + (-1.6) = (+1.6)$$

As can be seen, global emissions have been rising at a rate of about 1.6% per year, driven by the growth of population and of economic activity.

Within this generalization, countries vary widely. (Unless otherwise noted, comments about countries refer to the top-20 emitters as of 2005, who accounted for about 75% of world emissions that year.) Between 1990 and 2005, in some countries, including Brazil, Mexico, Indonesia, and South Africa, population growth alone exceeded the decline in intensity. For most countries, and for the world as a whole, per capita GDP growth exceeded the intensity improvement each achieved. Countries for whom intensity improvements were greater than their per capita GDP increases included Germany, the United Kingdom, the United States, France, and South Africa. And both the Russian Federation and the Ukraine, following their economic contractions in the 1990s, posted negative numbers for population, per capita income, intensity, and GHG emissions between 1990 and 2005. Besides the Russian Federation and the Ukraine, only the United Kingdom and Germany reduced their GHG emissions for the period (Germany being helped by reductions in the former East Germany).

Stabilizing greenhouse gas emissions would mean the rate of change equals zero. Globally, with a population growth rate of 1.4% per year and an income growth rate of 1.7% per year, intensity would have to decline at a rate of -3.1% per year to hold emissions at the level of the year that rate of decline went into effect. Within the United States, at the 1990-2005 population growth rate of 1.1% per year and income growth rate of 1.8% per year, intensity would have had to decline at a rate of -2.9% per year to hold emissions level; however, U.S. intensity declined at a rate of -1.9%, leaving emissions to grow at 1.0% per year.

Looking to the future, under auspices of the Copenhagen Accord, the United States has submitted a target of reducing emissions from the 2005 level by 17% in 2020. This would require the United States to reduce the intensity of its emissions by some -4.6% per year during the 2010-2020 decade. This implies that the rate of intensity decline needs to better than double.

INTRODUCTION

The interactions of three variables underlie debates on the issue of climate change[1] and what responses might be justified: the magnitude and rates of change of (1) population, (2) incomes, and (3) intensity of greenhouse gas emissions relative to economic activities. This report examines the interrelationships of the variables to explore their implications for policies that address climate change.

Both internationally and domestically, initiatives are underway both to better understand climate change and to take steps to slow, stop, and reverse the overall growth in greenhouse gas emissions,[2] the most important of which is carbon dioxide (CO2), emitted by the combustion of fossil fuels.

These initiatives include the following bulleted items.

- The United Nations Framework Convention on Climate Change (UNFCCC), to which the United States and almost all other nations are parties. Its stated objective is to stabilize greenhouse gas concentrations in the atmosphere at levels that "would prevent dangerous interference with the climate system."[3] It established the principle that all nations should take action, and that developed nations should take the lead in reducing emissions. It required parties to prepare national action plans to achieve reductions, with developed countries aiming to reduce year 2000 emissions to 1990 levels. It required preparation of inventories of emissions and annual reports. And it set up a process for the parties to continue meeting.
- The Kyoto Protocol, an agreement negotiated under the auspices of the UNFCCC; 169 nations—but *not* the United States—are parties to the protocol. Even as the Framework Convention was going into force, it was recognized that most nations would not meet their 2000 aims of holding emissions at 1990 levels. Ensuing negotiations, in which the United States participated, led to the Kyoto Protocol, which called for mandatory reductions in greenhouse gases for the period 2008-2012 by developed nations—but not by developing ones. With the United States not participating, and with developing nations not required to make GHG reductions (and China became the world's

largest emitter in about 2005), the reductions mandated by Kyoto will not significantly slow the accumulation of GHG in the atmosphere, although much is being learned of the institutional structures, such as the European Union's "cap and trade" mechanism, for handling reductions. Even as the Kyoto process began, negotiations on next steps resumed.

- The Asia-Pacific Partnership on Clean Development and Climate (APP) composed of the United States, China, India, Japan, Australia, and South Korea. The purposes of the partnership are to create a voluntary, non-legally binding framework for international cooperation to facilitate the development, diffusion, deployment, and transfer of existing, emerging, and longer-term cost-effective, cleaner, more efficient technologies and practices among the partners through concrete and substantial cooperation, so as to achieve practical results. It has the goal of meeting "national pollution reduction, energy security and climate change concerns, consistent with the principles of the U.N. Framework Convention on Climate Change (UNFCCC)."[4]
- The Copenhagen Accord is an agreement of the parties to the UNFCCC to begin establishing actions to follow on the Kyoto Protocol. The Copenhagen Accord[5] does not mandate specific reductions, but sets a goal of reducing global emissions "so as to hold the increase in global temperature below 2 degrees C, and take action to meet this objective consistent with science and on the basis of equity." Annex I nations commit to implement "quantified economy-wide emissions targets for 2020" and non-Annex I nations commit to implement "mitigation actions." Both sets of nations commit to reporting and verification procedures "in accordance with guidelines adopted by the Conference of the Parties." (Monitoring, reporting, and verification were a key demand of the United States of developing nations.) Also, the accord contained the promise that developed countries would make available $100 billion a year by 2020 "to address the needs of developing countries."

However, these several, related efforts have had to struggle with substantive economic, technical, and political differences among regional, national, and local circumstances. Foremost among these differences is the divide between developed and less-developed nations.[6] Conflict arises because any pressure to reduce emissions comes up against increases in emissions likely to result from energy use fueling economic development and raising standards of living in developing economies, which contain a large share of the world's population. Even in many developed nations, efforts to constrain emissions by, for example, conservation, increased energy efficiency, and use of energy sources that emit less or no CO_2, have been outstripped by increases in total energy use associated with population and economic growth. For example, between 1990 and 2005, in the United States, the greenhouse gas intensity of the economy declined at a rate of -1.9% per year, but total emissions increased at the rate of 1.0% per year.[7] Although some countries have experienced declines in emissions—either through economic contraction[8] or deliberate policies—the overall trend, both globally and for most individual nations, reflects increasing emissions.

This upward trend in greenhouse gas emissions runs counter to the long-term objectives of these climate change initiatives. This report identifies drivers of the increase in emissions and explores their implications for efforts to reduce emissions. During this exploration, it is useful to bear in mind that although short-term efforts may not achieve emissions reductions that immediately meet goals to prevent dangerous interference with the climate system, such endeavors may nevertheless establish a basis for longer-term efforts.

GREENHOUSE GAS EMISSION VARIABLES

The analysis below, which uses data from the World Resources Institute's Climate Analysis Indicators Tool (CAIT),[9] is based on the following relationships:

$$(Population) \times (per\ capita\ GDP) \times (Intensity_{ghg}) = Emissions_{ghg}$$

Equation 1

The CAIT database includes 185 nations (plus a separate entry for the European Union) with a 2005 population of 6.462 billion, compared with 191 members of the United Nations and with a 2005 world population count of 6.470 billion by the U.S. Census Bureau.[10] Average income is measured as per capita Gross Domestic Product (GDP), in international dollars of purchasing power parity ($PPP).[11] (Note that population times per capita GDP equals GDP.) Greenhouse gas intensity is measured as tons of emissions in carbon equivalents[12] per million dollars of GDP.

CHARACTERISTICS OF INTENSITY

Intensity can be expressed in many different ways; for example, as CO_2 emitted per million \$PPP, as all six greenhouse gases emitted per million \$PPP, and as CO_2 or greenhouse gases emitted per unit of some activity, such as electricity generated or vehicle miles traveled. Also, measures of intensity can include or exclude emissions or sequestration associated with land use changes.

In this analysis, intensity is identified as greenhouse gas (GHG) intensity (all six greenhouse gases of the UNFCCC) or as CO_2 intensity (referring only to CO_2 emissions from energy use and cement manufacture). In both cases, tonnage of emissions is expressed in carbon equivalents. CAIT has data on all six greenhouse gases only for 1990, 1995, 2000, and 2005; analyses referring to other years necessarily include only CO_2.

Unless otherwise specified, land use changes are not included in emissions or intensity data cited in this report.

Using international, purchasing power parity dollars can yield figures different from analyses using other economic measures, such as market exchange rate dollars. Intensity figures in this report, derived using \$PPP, may differ from comparable intensity figures in other studies using other GDP measures.

For the United States, using international \$PPP (or market exchange rate dollars) for GDP, CAIT yields a decline in intensity for U.S. emissions of all greenhouse gases between 1990 and 2005 of -1.9% per year.

To ensure consistency, CAIT emissions data and international \$PPP are used throughout this report, unless stated otherwise.

Table 1 provides a snapshot of the **equation 1** variables for the top 20 greenhouse gas emitters in the year 2005,[13] plus for the European Union 27, and for the world. The data reflect the wide range of circumstances faced by any initiative to address GHGs. However, it is the way those variables are changing that illuminates both the seemingly inexorable rise in GHG emissions and the challenge of reducing them. A variable changing at an annual rate of 6.9% doubles in 10 years; at an annual rate of 3%, it doubles in 23 years.

The growth rate of each of the variables of equation 1 can be expressed as an exponent, the annual percentage rate of change over some time period (see

Growth Relationship of Greenhouse Gas Drivers). As exponents of multiplicands are added, relationship among the variables can be simply expressed: the growth rates of the three variables on the left side of the equation sum to the growth rate of the variable (emissions) on the right side. Thus, if the sum of the three growth rate variables on the left is positive, emissions are rising; if the sum is negative, emissions are declining; and if the sum is zero, emissions are constant.

Table 1. Drivers of Greenhouse Gas Emissions: Top 20 Emitting Countries, 2005 (Excludes land use changes)

Country	Population (in 1,000s)	Per Capita GDP (2005 Int'l $PPP/ person)	Intensity (Tons Cequiv/ million 2005 Int'l $PPP)	Total GHG Emissions (MMTCE)
China	1,304,500	4,088	369.4	1,970.3
United States	296,507	41,813	153.3	1,900.6
EU-27	490,032	26,592	105.7	1,377.7
Russian Fed	143,150	11,861	315.1	534.9
India	1,094,583	2,230	207.2	505.7
Japan	127,773	30,290	94.7	366.5
Brazil	186,831	8,474	174.8	276.8
Germany	82,469	30,445	106.2	266.8
Canada	32,312	34,972	176.7	199.7
U.K.	60,226	31,371	92.4	174.6
Mexico	103,089	11,387	146.4	171.9
Indonesia	220,558	3,209	229.2	162.2
Iran	69,087	9,314	240.2	154.6
Italy	58,607	27,750	94.9	154.4
France	60,873	30,591	80.7	150.2
S. Korea	48,294	21,273	145.8	149.8
Australia	20,400	31,656	231.9	149.7
Ukraine	47,105	5,583	503.0	132.3
Spain	43,398	27,180	101.5	119.7
S. Africa	46,892	8,478	290.3	115.4
Turkey	72,065	10,370	143.6	107.3
WORLD	6,461,584	8,708	188.1	10,569.3

Source: Climate Analysis Indicators Tool (CAIT), version 7.0 (Washington, DC: World Resources Institute, 2010).

Growth Relationship of Greenhouse Gas Drivers

Incorporating growth, **equation 1** becomes

$$(Population)e^{k_p t} \times (percapitaGDP)e^{k_g t} \times (Intensity)e^{k_i t} = (Emissions)e^{k_e t}$$

Equation 2

in which k_p = population growth rate, k_g = per capita GDP growth rate, k_i = intensity growth rate, and k_e = emissions growth rate; t = time; and e = a constant 2.71828 (the base of natural logarithms).

The exponents of multiplicands are added, so

$$(k_p + k_g + k_i) = k_e$$

Equation 3

This growth relationship among the variables makes explicit why there is upward pressure on GHG emissions. For nearly all nations, population is increasing, with developing nations typically having the highest rate. Thus population growth rate is positive globally and for most nations; it is zero or negative for only a few nations.[14] The economic development of less-developed nations is a global objective acknowledged by the UNFCCC; developed nations also promote economic growth to raise living standards. Thus economies are growing globally and for most nations. With the population and economic activity variables positive, emissions will be rising unless the decline in intensity exceeds the growth in population and economic activity, which has seldom been the case. If the goal is to reduce GHG emissions, the larger the *negative* change in intensity, the better.

Table 2 shows the changes in these variables for 1990 - 2005. (The figures in the right-most column are taken from the CAIT database.[15]) As the table shows, global growth rates for population and per capita income out-paced the rate of decline in intensity—so GHG emissions rose; this is also true of the majority of nations, including the United States. Circumstances in several individual countries highlight some important points about GHG emissions and their potential control.

- First, for many nations, population growth is an important contributor to the increase in GHG emissions. For Brazil, Mexico, Indonesia,

Iran, Australia, Spain, South Africa, and Turkey, any improvements in intensity were annulled by increases in population alone.[16]

- Second, developing countries, focused on developing their economies, have increasing GHG emissions even when they manage to improve intensity (e.g., China, India, Mexico, Indonesia, South Korea, and South Africa). For these countries, population growth combined with per capita GDP growth overwhelmed whatever intensity improvements they achieved.[17]
- Third, lower emissions can be associated with decreasing economic activity. For the Russian Federation and Ukraine, economic contraction following the dissolution of the Soviet Union contributed to decreases in their emissions. During 2008-2009, GHG emission reductions occurred in many if not most nations as a result of the global recession, and almost certainly declined globally.
- Fourth, several developed countries improved per capita GDP while holding their GHG emissions to a 1% increase or less: the United States, Japan, Germany, the United Kingdom, Italy, and France. Germany and the United Kingdom (and also the European Union 27) actually decreased their emissions.
- Fifth, in some developed countries, income growth alone exceeded the decline in intensity (e.g., Japan, Italy, Canada, Australia, and Spain).

Stabilizing emissions would require an accelerated decline in intensity.[18] For global emissions to have met the UNFCCC voluntary goal of being at 1990 levels in 2000, intensity would have had to decline at the rate of -2.9% per year, rather than at the actual -2.0%.[19] For the United States, the situation was similar: for emissions in 2000 to have remained at 1990 levels, intensity would have had to decline at the rate of -3.2% per year, rather than the actual -1.9%.[20] Looking to the future, this relationship holds—absent a declining population or a contracting economy, GHG emissions can be expected to decline only if intensity declines at a rate faster than it has been.

How fast and how far might intensity be driven down? There are two ways to approach this question: one is to examine the sources of emissions and consider how much and how fast they could be curtailed; a second is to assess what level of greenhouse gases can be emitted to the atmosphere without causing "dangerous interference with the climate system" (in the words of the UNFCCC) and to calculate from those emissions what the intensity would have to be over time, taking into account population and income growth.

Table 2. Annual Percentage Rate of Change in Factors Affecting Greenhouse Gas Emissions: Top 20 Emitting Countries, 1990-2005 (Excludes land use changes)

Country	Population (average annual %)	Per Capita GDP (average annual %)	Intensity (average annual %)	Total GHG Emissions (average annual %)
China	0.9	9.1	-4.9	4.8[a]
United States	1.1	1.8	-1.9	1.0
EU-27	0.3	1.8	-2.4	-0.4
Russian Fed	-0.2	-0.4	-2.1	-2.7
India	1.7	4.2	-2.3	3.5
Japan	0.2	1.0	-0.4	0.9
Brazil	1.5	1.1	0.0	2.6
Germany	0.3	1.3	-2.9	-1.3
Canada	1.0	1.8	-1.2	1.6
U.K.	0.3	2.1	-3.0	-0.6
Mexico	1.4	1.5	-0.6	2.3
Indonesia	1.4	2.9	-0.5	3.8
Iran	1.6	2.7	1.2	5.6
Italy	0.2	1.1	-0.6	0.8
France	0.5	1.4	-1.7	0.1
S. Korea	0.8	4.7	-1.3	4.2
Australia	1.2	2.0	-1.0	2.2
Ukraine	-0.6	-2.4	-1.2	-4.2
Spain	0.7	2.2	0.0	3.0
S. Africa	1.9	0.6	-0.9	1.6
Turkey	1.6	2.3	-1.1	2.7
WORLD	1.4	1.7	-1.6	1.5

Source: Climate analysis Indicators Tool (CAIT), version 7.0 (Washington, DC: World Resources Institute, 2010).

a. These figures are taken from the CAIT data base. In principle, they could be calculated by adding the three left-hand data columns; in fact, a number of rows do not add; this may be due to rounding or, where discrepancies are large, from inconsistencies in the underlying reported data.

Table 3. GHG Emissions by Sector:Top 20 Emitting Countries, 2005 (Includes International Bunkers and Land-Use Change and Forestry)

Country	Energy (CO_2, N_2O, and CH_4) MMTCE	Industrial Processes (All 6 GHG) MMTCE	Agriculture (N_2O and CH_4) MMTCE	Waste (e.g. landfills) (N_2O and CH_4) MMTCE	Land-Use Change & Forestry (CO_2) MMTCE	International Bunkers (CO_2) MMTCE	Total (All 6 GHG) MMTCE
China	1,441	183[a]	304	48	-13	8	1,970
United States	1,652	68	121	51	-32	36	1,896
EU-27	1,134	71	137	36	—[b]	77	1,456
Russian Fed	474	12	32	13	16	4	552
India	342	24	110	34	—	3	512[c]
Japan	337	21	10	3	—	11	381[c]
Brazil	94	9	161	12	500	4	780
Germany	229	10	23	4	—	8	274[c]
Canada	168	6a	20	7	18	1	221
U.K.	154	6	13	3	—	11	187[c]
Mexico	134	8	21	13	11	3	189
Indonesia	108	5	36	9	398	1	558
Iran	132	5	10	4	—	1	153[c]
Italy	129	10	11	3	—	5	159[c]
France	109	9	28	3	—	7	156[c]
S. Korea	130	16	5	4	—	11	166[c]
Australia	116	3	30	3	—	3	156[c]
Ukraine	116	3	12	4	—	0	135[c]
Spain	94	9	12	3	—	9	128[c]
S. Africa	94	4	11	6	—	3	118[c]
Turkey	74	6[a]	21	5	9	2	118
WORLD	7,761	514	1,658	387	1,467	257	12,044

Source: Climate analysis Indicators Tool (CAIT), version 7.0 (Washington, DC: World Resources Institute, 2010).

a. CH_4 data not available.

b. Not available

c. Data from Land-Use Change & Forestry not available.

Sectoral Breakdown of GHG Emissions

Greenhouse gas emissions result from diverse human activities, including agriculture and the combustion of fossil fuels – the latter providing the energy that has driven the industrial revolution and accounting for much of the rise in CO2 levels in recent centuries. This section of the report examines several of those major sources of emissions.

Table 3 presents emissions data by economic sector for the top 20 emitting nations (plus the EU27 and the world), including Land-Use Change and Forestry, and International Bunkers (so the total is different than in **Table 1**). As **Table 3** shows, the energy sector is by far the largest contributor of greenhouse gases, accounting for 64% of total world emissions in 2005; the agricultural sector is second, accounting for about 14%. These two sectors dominate for almost all countries (industrial process emissions rank second for Japan and South Korea) – except for the dominance of Land-Use Change and Forestry for Brazil and Indonesia.

Table 4 presents a breakdown of the energy sector emissions. Electricity and heat contributes the largest share, accounting for about 43% of total energy sector emissions in 2005, followed by transportation at about 19%, manufacturing at about 18%, other fuel combustion at about 13%, and fugitive emissions at about 6%.

The most revealing aspect of sectoral emissions emerges from **Table 5**, which shows their rates of change, and **Table 6**, which shows the rates of change of the energy subsectors.[21]

Global emissions are growing fastest in the Industrial Processes sector[22] (4.0%/year); next is the International Bunkers sector, growing at 2.9%/year: but as these two sectors are much smaller shares of total emissions than energy (see **Table 3**), the increases are relatively small in absolute terms; however, the rate of increase is substantial for nations that are industrializing, especially China, India, and South Korea.

The largest absolute increase in emissions is driven by the rate of increase for the energy sector, growing at 1.6%/year. Within that sector (see **Table 6**), the most rapidly growing subsector is electricity and heat energy, at 2.5% per year, led by developing nations, especially China, India, Brazil, South Korea, Iran, and Indonesia, and also by Spain and Turkey. In contrast, for the EU27, the rate and absolute emissions for the subsector declined slightly; but for the Russian Federation and Ukraine, the rate and absolute emissions declined substantially as their economies contracted. The next fastest growing subsector is transportation, at 2.0% a year, with every nation showing a positive rate of growth except the Russian Federation and Ukraine, with their contracting economies during the 1990s, and Germany, with a minimal decrease. The fastest rates of transportation emissions growth occurred in China, Iran, Indonesia, and South Korea.

Table 4. Energy Sector GHG Emissions: Top 20 Emitting Countries, 2005 (Excludes land use changes)

Country	Electricity and Heat (CO_2) MMTCE	Manufacture and Construction (CO_2) MMTCE	Transportation (CO_2) MMTCE	Other Fuel Combustion (CO_2, N_2O & CH4) MMTCE	Fugitive Emissions (CO_2 & CH_4) MMTCE
China	728	435	91	148	39
United States	746	171	493	185	57
EU-27	445	180	256	228	25
Russian Fed	255	61	60	45	54
India	189	69	27	44	13
Japan	139	78	68	52	0
Brazil	16	26	37	12	3
Germany	100	31	42	51	4
Canada	52	28	44	34	12
U.K.	65	18	35	32	5
Mexico	49	16	36	11	23
Indonesia	34	25	20	13	16
Iran	29	19	28	34	23
Italy	45	22	32	28	2
France	20	20	36	32	1
S. Korea	60	26	24	19	1
Australia	66	13	22	8	8
Ukraine	37	25	8	14	32
Spain	35	18	30	11	1
S. Africa	58	14	12	8	2
Turkey	22	16	10	13	14
WORLD	3,367	1,427	1,465	1,024	477

Source: Climate analysis Indicators Tool (CAIT), version 7.0 (Washington, DC: World Resources Institute, 2010).

Table 5. Annual Percentage Rate of Change of GHG Emissions by Sector: Top 20 Emitting Countries, 1990-2005 (Includes International Bunkers and Land-Use Change and Forestry)

Country	Energy (CO_2, N_2O, and CH_4) %	Industrial Processes (All 6 GHG) %	Agriculture (N_2O and CH_4) %	Waste (e.g. landfills) (N_2O and CH_4) %	Land-Use Change & Forestry (CO_2) %	International Bunkers (CO_2) %	Total (All 6 GHG) %
China	5.4	11.4[a]	1.4	0.9	[b]	12.7	4.3
United States	1.0	3.0	0.2	-0.9		0.2	1.0
EU-27	-0.2	-0.8	-1.2	-2.5		3.2	-0.3[d]
Russian Fed	-2.6	-2.6	-4.1	-0.6		-4.9	-2.6
India	4.4	7.5	1.3	1.8		[c]	3.6[d]
Japan	0.9	1.1	-0.7	1.0		1.9	0.9[d]
Brazil	3.5	2.4	2.2	1.4			0.8
Germany	-1.1	-1.6	-1.8	-6.7		2.0	-1.3[d]
Canada	1.7	-1.5[a]	1.6	2.0			1.2
U.K.	-0.3	-4.4	-0.7	-5.8		3.7	-0.5[d]
Mexico	2.5	4.6	0.9	1.6			2.1
Indonesia	5.0	6.4	1.5	1.4			0.9
Iran	6.0	6.0	3.2	1.6			5.6[d]
Italy	0.9	1.0	-0.2	-1.1		2.9	0.8[d]
France	0.5	-1.9	-0.4	-1.0		2.5	0.2[d]
S. Korea	4.8	5.6	-0.0	-3.6		13.4	4.5[d]
Australia	2.7	1.4	0.9	1.0			2.2[d]
Ukraine	-4.4	-4.0	-3.8	-0.7			-4.2[d]
Spain	3.3	2.4	1.1	3.5		5.7	3.2[d]
S. Africa	1.7	5.6	-0.3	1.2			1.6[d]
Turkey	3.9	4.0[d]	-0.3	1.6			2.6
WORLD	1.6	4.0	1.0	0.6	-0.8	2.9	1.2

Source: Climate analysis Indicators Tool (CAIT), version 7.0 (Washington, DC: World Resources Institute, 2010).

a. CH_4 data not available.

b. Individual country data unavailable or insufficient for useful analysis of trends.

c. No entry indicates not calculated for countries for which both 1990 and 2005 emissions were <5 MMTCE.

d. Data from Land-Use and Forestry not available.

Table 6. Annual Percentage Rate of Change of Energy Sector GHG Emissions: Top 20 Emitting Countries, 1990-2005

Country	Electricity and Heat (CO_2) %	Manufacture and Construction (CO_2) %	Transportation (CO_2) %	Other Fuel Combustion (CO_2, N_2O & CH_4) %	Fugitive Emissions (CO_2 & CH_4) %
China	9.2	3.9	7.2	0.0	0.7
United States	1.6	-0.7	1.6	0.0	-1.3
EU-27	-0.2	-1.5	1.5	-0.3	-2.9
Russian Fed	-1.8	-1.8	-2.0	-5.5	-4.6
India	6.7	2.8	1.2	2.5	3.3
Japan	1.5	-0.2	1.1	1.0	[a]
Brazil	5.2	3.6	3.6	1.4	
Germany	-0.6	-2.9	-0.1	-1.2	—[b]
Canada	2.0	1.1	3.6	1.9	1.9
U.K.	-0.2	-1.7	0.6	0.0	-3.2
Mexico	3.7	-1.5	2.8	1.3	4.5
Indonesia	6.0	7.2	5.8	2.9	2.1
Iran	6.7	3.2	6.6	5.5	8.7
Italy	1.0	-0.3	1.5	1.2	
France	0.9	-0.7	1.1	0.8	
S. Korea	8.7	3.9	4.7	0.0	
Australia	3.7	0.1	1.7	3.5	1.7
Ukraine	-6.2	-5.2	-3.8	-3.5	-0.9
Spain	3.5	2.4	3.7	3.9	
S. Africa	2.6	-1.9	2.6	3.6	
Turkey	5.0	3.5	2.0	2.8	5.9
WORLD	2.5	1.0	2.0	-0.0	0.2

Source: Climate analysis Indicators Tool (CAIT), version 7.0 (Washington, DC: World Resources Institute, 2010).

a. No entry indicates not calculated for countries for which both 1990 and 2005 emissions were <5 MMTCE.

b. Not available.

Energy Use as a CO_2 Intensity Driver

The previous section looked at emissions and the rate of change, 1990-2005, for all six greenhouse gases and all sectors of the economy (insofar as data are available). Of the six greenhouse gases, CO_2 dominates, accounting for 76.8% of the carbon equivalents of global GHG emissions in 2005 and 84.6% of U.S. GHG emissions (these figures include Land-Use Change and Forestry, and International Bunkers). Overwhelmingly, the source of that CO_2 is energy use: for world CO_2 emissions, energy use accounts for 77.9%; for the United States, energy use accounts for 98.8%.

Two factors largely determine the intensity of CO2 emissions of a nation's economy: energy intensity (energy per unit of GDP) and the fuel mix (emissions per unit of energy):[23]

$$\frac{Energy\ Use}{GDP} \times \frac{Emissions_{co2}}{Energy\ Use} = \frac{Emissions_{co2}}{GDP}$$ **Equation 4**

Table 7 presents data on energy sector CO_2 emissions for 2006. The first data column represents energy intensity of the economy, measured in 1,000 tonnes of oil equivalent (toe) per million $PPP. The smaller the number, the more efficiently energy is used to support economic activity in that country. For the world, the energy intensity of the global economy is 0.19; of the top-20 emitting nations, 13 equal or better the world average. Seven countries, Japan, Mexico, Germany, the United Kingdom, Italy, Spain, and Turkey equal or better the efficiency of the EU-27, at 0.13; China, the Russian Federation, Ukraine, and South Africa are the least efficient, at 0.31 or worse.

In general, the higher the number in column one, the more least-cost options that nation should be able to find for reducing energy use without adversely affecting the overall economy. Improvements could come, for example, from upgrading boilers, substituting gas-combined cycle electricity generation, improving the efficiency of the electricity grid, or upping the efficiency of the vehicle fleet.

The second data column in **Table 7** reflects the carbon content of the mix of fuels comprising energy use, measured as tons of carbon (C) per 1,000 tonnes of oil equivalent. The world average is 670. Of the top-20 GHG emitters, China has the highest emissions for the energy it uses, at 900 tons of carbon per 1000 tonnes of oil equivalent; France—with nuclear power dominating its electricity generating sector—is lowest, at 390. The United States is just over the world average, at 680.

The smaller the number, the less CO_2 being emitted by the energy used. Higher numbers would generally reflect a higher proportion of coal combusted in the electricity-generating, manufacturing, and heating sectors and a low transportation fleet fuel economy; lower numbers would generally reflect a higher proportion of hydropower, renewables, or nuclear power in the electricity, manufacturing, and heating sector, and a high transportation fleet fuel economy. Again, in many cases, the higher the number, the more least-cost options for lowering CO_2 emissions without adversely affecting the overall economy, for example by substituting natural gas for coal or renewables for oil.

Table 7. CO_2 Emissions Intensity of the Energy Sector: Top 20 Emitting Countries, 2006

Country	Energy Intensity (1,000 toe/ million 2005 $PPP)	CO_2 Intensity of Energy Sector (Tons C/1,000 toe)	CO_2 Intensity of Economy (Tons C/ million 2005 $PPP)	Total CO_2 Emissions from Energy Use (MMTCE)
China	0.32	900	288	1,530
United States	0.19	680	129	1,561
EU-27	0.14	620	87	1,089
Russian Fed	0.38	650	247	433
India	0.22	640	141	341
Japan	0.14	650	91	331
Brazil	0.13	430	56	92
Germany	0.14	660	92	225
Canada	0.24	560	134	148
U.K.	0.12	640	77	147
Mexico	0.15	680	102	115
Indonesia	0.25	550	138	94
Iran	0.25	750	188	124
Italy	0.11	700	77	122
France	0.15	390	58	103
S. Korea	0.21	630	132	130
Australia	0.19	890	169	108
Ukraine	0.54	630	340	85
Spain	0.12	670	80	89
S. Africa	0.32	730	234	93
Turkey	0.11	760	84	65
WORLD	0.20	670	134	7,429

Source: Climate analysis Indicators Tool (CAIT), version 7.0 (Washington, DC: World Resources Institute, 2010). CRS calculations.

The third data column in **Table 7** contains each nation's intensity of carbon emissions for the energy sector; it is the product of the first and second data columns. (Note that this intensity number is for CO_2 emissions only, and is thus different from *greenhouse gas intensity*, which includes CO_2 plus five other gases.) The higher the number, the less efficiently the economy is using carbon-emitting energy. The highest intensity nations are Ukraine, China, the Russian Federation, and South Africa; the lowest are France and Brazil, meaning that they get the most economic output for the emissions of CO_2 from the energy they use. The United States at 130 is slightly more efficient that the world average of 138, but less efficient than, for example, the European Union-27, at 86, and Japan and Turkey, both at 88.

The last column in the table provides data on total emissions from energy use for 2006[24] – a year later than the same data series in column one in **Table 3.**

Another question is the relationship between new economic growth and emissions, which are often influenced by the degree of industrialization and the prices and availability of different fuels. **Table 8** compares this by providing information on the annual rates of change of factors affecting CO_2 emissions from energy use. The first three data columns parallel the first three in **Table 7**, giving the rates of change during 1990-2005. In terms of CO_2 emissions, negative numbers mean that over time a nation is getting more economic activity for less energy (first data column) and more energy for less CO_2 (second data column). As **Table 8** shows, there are wide variations among nations. For example, China's economy made rapid progress in using energy more efficiently (energy intensity of -5.1% per year), even though the energy it used actually produced more CO_2 per unit of energy (+1.3% per year). A number of countries, including the EU-27, improved efficiency and reduced emissions per unit of energy used. The third data column, which should be the sum of the first two, is negative if, overall, the country is producing more economic activity for theCO_2 emitted. The fourth and fifth columns in Table 8 give the rates of change of the nations' GDPs and total CO_2 emissions from energy use. A nation's rate of change of CO_2 intensity can be negative, but if GDP is growing faster than CO_2 intensity is declining, emissions will rise (the last column).[25]

The carbon intensity of energy use—that is, the consequences of fuel mix—is especially notable in looking at the energy mix of electricity generation, as discussed in the next section.

Table 8. Annual Percentage Rate of Change in Factors Affecting CO_2 Emissions from Energy Use: Top 20 Emitting Countries, 1990-2005

Country	Energy Intensity %	CO_2 Intensity of Energy Used %	Energy Sector CO_2 Intensity %	Per Capita GDP %	Total CO_2 Emissions of Energy Use %
China	-4.8	1.3	-3.5	9.2	6.0
United States	-1.4	-0.2	-1.6	1.8	1.0
EU-27	-1.2	-0.7	-1.8a	1.9	-0.1
Russian Fed	-1.3	-0.4	-1.7	0.1	-2.0
India	-2.1	1.2	-0.9	4.4	4.7
Japan	0.0	-0.3	-0.3	1.1	0.8
Brazil	0.0	0.3	0.3	1.2	3.4
Germany	-1.2	-0.7	-2.0	1.4	-0.9
Canada	-1.0	-0.2	-1.2	1.8	1.4
U.K.	-1.8	-0.7	-2.5	2.1	-0.2
Mexico	0.0	0.0	0.0	1.6	2.3
Indonesia	-0.7	1.8	1.1	3.0	5.3
Iran	1.4	0.2	1.6	2.8	6.1
Italy	0.0	-0.6	-0.6	1.1	0.7
France	-0.4	-0.8	-1.2	1.4	0.4
S. Korea	0.6	-0.8	-0.2	4.7	4.7
Australia	-0.9	0.5	-0.4	2.0	2.6
Ukraine	-0.8	-1.1	-1.9	-1.8	-4.9
Spain	0.0	0.1	0.1	2.2	2.9
S. Africa	-0.2	-0.3	-0.5	0.8	1.9
Turkey	-0.5	0.3	-0.1	2.5	4.1
WORLD	-1.4	0.0	-1.4	1.8	1.8

Source: Climate Analysis Indicators Tool (CAIT), version 7.0 (Washington, DC: World Resources Institute, 2010). CRS calculations.

a. When the third column is not the sum of the first two, the discrepancy reflects rounding or, possibly, data shortcomings.

Carbon Intensity of Electricity Generation

Variations among countries of the carbon intensity of energy use (see **Table 5**) are strongly affected by the carbon intensity of electricity generation

– in the United States electricity generation accounts for 41% of total CO_2 emissions.[26] Differences among countries are marked, as depicted in **Table 9**.

Choices among generating technologies are the primary driver of disparities among countries in the carbon intensity of their electricity generation. In general, countries with high numbers generate a substantial proportion of their electricity by burning fossil fuels, and countries with low numbers generate large quantities of electricity by nuclear facilities, hydropower, or other renewables. For example, France, with the lowest carbon intensity of electricity production of 20.3 in 2006, generated 79.2% of its electricity by nuclear power, 11.7% by hydropower and other renewables, and about 9.6% by conventional thermal. The United States, with a carbon intensity of electricity production in 2006 of 148, generated 19.4% of its electricity by nuclear power, 9.8% by hydropower and other renewables, and about 71% by conventional thermal.[27]

Although a nation's electricity-generating technologies are obviously affected by its resource endowments in terms of hydropower and fossil fuels, choices can be made, as exemplified by France. In 1980, France's electricity was generated 27% by hydropower, 24% by nuclear, 27% by coal, and 19% by oil. By 1990, with electricity production up over 60%, nuclear had risen to a 75% share, whereas coal and oil had fallen to 8% and 2% shares, respect-tively.[28] Not only did nuclear power account for all the growth in electricity generation during the period, but it displaced half the coal-fired and more than three-quarters of the oil-fired electricity generation. In 1990, the electricity produced by nuclear power exceeded France's total amount of electricity generated 10 years earlier.

France's transition to nuclear power meant that its CO_2 intensity (i.e., CO_2 emissions/GDP) declined between 1980 and 1990 at a rate of -4.9% per year, and CO_2 emissions declined at a rate of -2.6% per year. Thus, between 1980 and 1990, France's total CO_2 emissions declined by 23%—at the same time its per capita GDP was growing by 20.4% (+1.9% per year).[29] Thus **equation 3** yields a negative growth in emissions (numbers do not add precisely, due to rounding):

France: CO_2 Emission Drivers, 1980-1990 (Annual rate of change, %)						
Population		Per Capita GDP		CO2 Intensity		CO2 Emissions
(0.5)	+	(1.9)	+	(-4.9)	=	(-2.6)

During the 1990s, the United Kingdom made a major shift from coal to natural gas in the generation of its electricity. In 1990, the United Kingdom's electricity was generated 21% by nuclear, 1% by natural gas, and 65% by coal. In 2000, with electricity generation up 17%, nuclear's share was 23%, whereas coal's share had dropped to 33% and natural gas's share had risen to 39%.[30] Because natural gas produces less total CO_2 per kilowatt hour than coal (at a ratio of about 0.6 to 1 on a Btu basis[31]), CO_2 intensity in the United Kingdom declined between 1990 and 2000 at a rate of -2.8% per year, and CO_2 emissions declined at a rate of -0.5% per year. Thus, between 1990 and 2000, total CO_2 emissions in the United Kingdom declined by 4.5% (-0.5% per year)—at the same time its per capita GDP was growing by 23.6% (+2.1% per year).[32] Thus **equation 3** yields a negative growth in emissions (numbers do not add precisely, due to rounding):

United Kingdom: CO_2 Emission Drivers, 1990-2000 (Annual rate of change, %)						
Population		Per Capita GDP		CO_2 Intensity		CO_2 Emissions
(0.3)	+	(2.1)	+	(-2.8)	=	(-0.5)

The examples of France and the United Kingdom show that for a period of time, at least, greenhouse gas intensity improvements can be sufficient to absorb growth in population and economic activity, so that actual emissions decline. The examples also show that the introduction of new technology can cause sudden shifts in emission rates.

The United States has also had periods when its CO_2 emissions declined. From 1980-1986, U.S. CO_2 intensity declined at a rate of -3.6% per year, and emissions declined at a rate of -0.5% per year. But after 1986 the rate of intensity decrease slowed: between 1987 and 2003, the intensity rate averaged about -1.7% per year. After 2003, through 2006 (the last year of CAIT's data), the rate of intensity decrease speeded up to an average annual -2.6%. Nonetheless, throughout the 1987-2006 period, the decrease failed to compensate for population and per capita GDP growth, so CO_2 emissions rose at 1.1% per year.[33] Over the longer term, therefore, emissions have risen: in terms of **equation 3**, U.S. CO_2 emissions for 1980-2005 are as follows (numbers do not add precisely, due to rounding):

United States: CO_2 Intensity, 1980-2005						
Population		Per Capita GDP		CO_2 Intensity		CO_2 Emissions
(1.1)	+	(2.0)	+	(-2.2)	=	(0.8)

Table 9. Carbon Intensity of Electricity Generation: Top 20 Emitting Countries, 2006

Country	Intensity (gC/kWh)
China	230.5
United States	148.4
EU-27	96.6
Russian Fed	a
India	254.6
Japan	113.6
Brazil	22.3
Germany	134.3
Canada	49.6
U.K.	135.5
Mexico	147.7
Indonesia	184.7
Iran	146.5
Italy	113.6
France	20.3
S. Korea	127.5
Country	**Intensity (gC/kWh)**
Australia	244.0
Ukraine	118.1
Spain	92.4
S. Africa	238.2
Turkey	122.7
WORLD	143.7

Source: Climate analysis Indicators Tool (CAIT), version 7.0 (Washington, DC: World Resources Institute, 2010).

a. Not available.

Carbon Intensity of Travel

The transportation sector is one of the fast-growing sources of emissions (see **Table 6**) – and it is proving one of the most intractable to reducing greenhouse gas emissions. Studies indicate that nations vary considerably in the energy efficiency and greenhouse gas emissions intensity of their transport sectors, but data are limited for making inter-country comparisons of the carbon intensity of passenger miles or of ton-miles. For example, one effort examining vehicle miles shows substantial variations among several nations, with the United States being the highest emitter per passenger vehicle[34] (**Figure 1**). To some extent, these variations reflect differing geographic, cultural, and infrastructure circumstances among the nations; however, as with the carbon intensity of electricity generation, a substantial cause of the variations is deliberate policies, such as fuel efficiency standards, emission standards, fuel taxes, and choices of investments in transportation infrastructure.

For the United States, the Bureau of Transportation Statistics provides data on the energy intensity of passenger modes (**Table 10**).

Two important points emerge from **Table 10**. First, transportation efficiency for several modes has improved over time. Air traffic gained efficiency in the transition to jets and larger aircraft. Vehicular passenger miles have gained efficiency, but at a slowing pace. On the other hand, transit motor bus efficiency per passenger mile has gone up and down. Second, the choice of transportation mode, which can be affected by infrastructure investments and other public policies,[35] substantively affects passenger-mile efficiency. Amtrak and, by extension, commuter rail, is considerably more efficient than any of the other choices, except motorcycles. Moreover, within the highway mode, efficiency varies significantly: in 2000, passenger cars were 20% more efficient on average than pickups, SUVs, and minivans, but in 2006 improvements in the latter had reduced the difference to 12%.

Table 10. Energy Intensity of Passenger Modes: United States, 1970-2006 (Btus per passenger-mile)

Passenger Modes	1970	1980	1990	2000	2005	2006
Air, certified carrier						
Domestic	10,185	5,742	4,932	3,883	3,222	3,098
International	10,986	4,339	4,546	3,833	3,813	3,691
Highway						
Passenger car	4,841	4,348	3,811	3,589	3,585	3,525
Passenger Modes	**1970**	**1980**	**1990**	**2000**	**2005**	**2006**
Pickup, SUV, minivan	6,810	5,709	4,539	4,509	4,077	4,016
Motorcycle	2,500	2,125	2,227	2,273	1,784	1,754
Transit motor bus	—	2,742	3,723	4,147	3,393	3,262
Amtrak	—	2,148	2,066	2,134	—	—

Source: Bureau of Transportation Statistics: http://www.bts.gov/publications/national_transportation_statistics/ html/table_04_20.html.

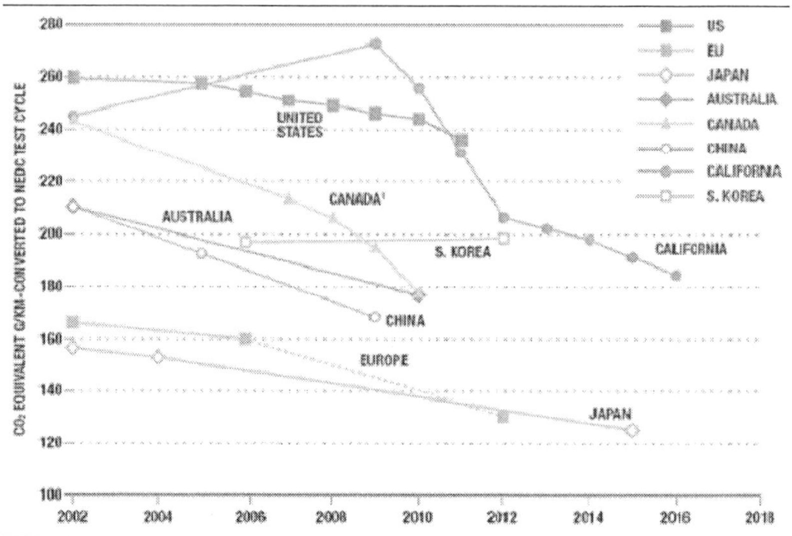

Source: Feng An, et al., *Passenger Vehicle Greenhouse Gas and Fuel Economy Standards: A Global Update*, International Council on Clean Transportation (July 2007), p. 8.

Note: Solid lines denote actual performance or projected performance due to adopted regulations; dotted lines denote proposed standards; Values normalized to NEDC last cycle in grams of CO2-equivalent per km.

1. For Canada, the program includes in-use vehicles. The resulting uncertainty on new vehicle fuel economy was not quantified.

Figure 1. Actual and Projected GHG Emission for New Passenger Vehicles by Country, 2002-2018.

All in all, it appears that policy choices can affect the energy intensity of travel, and thus opportunities for improvement exist. Because there is clearly a limit on greenhouse gas emission reductions to be achieved by heightened efficiencies in the transportation sector, interest turns to alternative fuels that do not generate greenhouse gases, including renewables and hydrogen. Brazil has made considerable progress in substituting ethanol for gasoline (40% by volume); however, the U.S. promotion of ethanol is still a small proportion of gasoline consumption (3.6% by volume in 2006), and there are questions about the net impact of ethanol use on CO2 emissions.[36] Hydrogen remains a distant possibility.

Effects of Land Use on Intensity

Land use changes can affect emissions (**Table 3**) and intensity (**Table 11**). They have been excluded from most analyses in this report because the data are limited and less robust than most of the emissions data, and because for most nations, taking it into account changes little.

Taking into account land use changes and forestry is highly time dependent: as nations clear land, develop their agriculture, and harvest forest resources, their emissions will rise; but for many nations, these activities occurred long ago. Only for those nations now at the point in their development where agricultural land clearing and logging are important activities, do substantial emissions result. Today, this is most notable in Indonesia and Brazil. In 2005, their emissions attributable to land use changes and forestry practices accounted for 71% and 64%, respectively, of their total GHG emissions. And as can be seen in **Table 11**, incorporating land use changes and forestry greatly increases the greenhouse gas intensity of their economies.

Even though land use changes may have a small effect on emissions for most countries, and the data lack robustness, including it in analyses can identify those situations where it is undeniably important and for which interventions might pay large dividends in terms of curtailing greenhouse gas emissions or sequestering CO_2.

Cumulative Emissions

That economic growth is a major underlying cause of the rise in greenhouse gas emissions is evident when one examines cumulative emissions: nations that achieved economic growth in the 19th and 20th Centuries account for the majority of the emissions over time. Moreover, as greenhouse gas emissions are long-lived in the atmosphere, their effect cumulates over time. A justification for developed nations taking the lead on reducing emissions, while giving developing ones the opportunity to increase emissions from activities that are necessary for economic development, is not just that developed nations are wealthier but also that they account for the bulk of cumulative emissions affecting climate. Data to assess cumulative emissions are limited. In general, data are available only for CO2 and are calculated from fuel use estimates; land use changes over long time spans are important, but data are scanty or unavailable. CAIT provides figures for CO2 emissions from fuel use, only from 1850, and not including land use changes (**Table 12**).

Table 11. Land Use Changes: Impact on Intensity of Greenhouse Gas Emissions— Top 20 Emitting Countries

Country	Intensity 2005 (excluding land use) tCeq/ million $PPP (all 6 GHG)	Intensity 2005 (including land use) tCeq/ million $PPP (all 6 GHG)	Intensity difference, with land use minus without land use	% difference
China	372	369	-3	-1
United States	153	150	-3	-2
EU-27	105	105	0	0
Russian Fed	313	322	9	3
India	208	—[a]	—	—
Japan	96	—	—	—
Brazil	174	490	316	182
Germany	103	—	—	—
Canada	178	194	16	9
U.K.	91	—	—	—
Mexico	136	144	8	4
Indonesia	226	790	564	250
Iran	236	—	—	—
Italy	93	—	—	—
France	80	—	—	—
S. Korea	142	—	—	—
Australia	236	—	—	—
Ukraine	512	—	—	—
Spain	100	—	—	—
S. Africa	289	—	—	—
Turkey	136	148	8	6
WORLD	182	208	26	14

Source: Climate analysis Indicators Tool (CAIT), version 7.0 (Washington, DC: World Resources Institute, 2010).

a. Not available.

Table 12. Cumulative CO2 Emissions from Energy: Top 20 Emitting Countries, 1850-2006 (Excludes Land Use Changes)

Country	Cumulative Emissions (MMTCE)	Percentage of World	Rank in World
China	27,075	8.6	2
United States	91,088	29.0	1
EU-27	83,447	26.6	
Russian Fed	25,404	8.1	3
India	7,487	2.4	8
Japan	12,155	3.9	6
Brazil	2,581	0.8	21
Germany	21,937	7.0	4
Canada	6,860	2.2	9
U.K.	18,623	5.9	5
Mexico	3,212	1.0	15
Indonesia	1,788	0.6	25
Iran	2,211	0.7	23
Italy	5,132	1.6	12
France	8,810	2.8	7
S. Korea	2,699	0.9	20
Australia	3,470	1.1	14
Ukraine	6,856	2.2	10
Spain	2,915	0.9	17
S. Africa	3,492	1.1	13
Turkey	1,490	0.5	29
WORLD	314,056	100.0	

Source: Climate analysis Indicators Tool (CAIT), version 7.0 (Washington, DC: World Resources Institute, 2010).

Because climate-forcing depends on the cumulative emissions, not current emissions, it is easy to see from **Table 12** why developing nations feel that developed ones should take the lead. Given CAIT data, the United States and the European Union-27 account for over half the cumulative CO2 emissions from energy use since 1850.

The data on cumulative emissions and on including or excluding land use changes (see **Table 11**) highlight why individual nations are so differently affected by proposals to reduce greenhouse gas emissions. Setting a baseline year for determining a nation's emissions means that countries that developed early could do so with no restrictions on the use of fuels and other resources regardless of their potential impact on climate, while those nations just now undergoing development might face restrictions. The emissions of already developed nations are embedded in their baselines. Similarly, whether certain activities such as land use changes are included or not affects what is in the baseline. The greenhouse gas emissions of Brazil and Indonesia, for example, increase markedly when emissions from land use changes of the last few decades are counted; but comparable land use changes in many other countries (e.g., the United States) happened in earlier centuries, and the resulting emissions count only toward cumulation, not against any current baseline.

INTERACTIONS OF THE VARIABLES

Numerous subtle and indirect interactions occur among population, income, intensity, energy use, land use changes, and emissions. These interactions affect policy choices concerning climate change because of their implications for other important social policy initiatives and objectives— most importantly, policies to promote economic development and income growth. These interactions also make difficult the projection of trends over time.

Economic development and growing incomes interact with population growth in two ways. First, birth rates tend to decline as incomes rise,[37] reducing one of the upward pressures on emissions. Most high-income nations have annual birth rates of 0.5% or lower, compared with developing nations with birth rates that in some cases exceed 2% per year. Second, the economic opportunity that many developed nations offer means they may have relatively high immigration rates, so their population growth is higher than their birth rate.[38] Overall, most demographers expect the rate of population growth to slow, although world population is projected to exceed 9 billion in 2050, with most of the increase in the developing world.[39]

Economic development and energy use are closely intertwined. The substitution of fossil fuel energy for human and animal power has been an important driver of the industrial revolution and consequent higher incomes. Indeed, for many, industrialization is synonymous with economic development. Yet at some point in development, the growth in incomes becomes at least partially detached from energy use, as energy costs lead to attention to energy efficiencies and as economies shift toward post-industrial services. Public policies can affect the relationship between economic development and growth and energy use in many ways, including taxation, infrastructure development, and research and development. The UNFCCC assumes that developing nations will inevitably have to exploit more energy as they give priority to reducing poverty. A key element of the climate change debate is how to decouple that economic development-energy use link.[40]

Income and emissions are related in another way, as well. In general, low-income people in developing nations focus their efforts on survival, whereas nations and individuals with higher incomes are likely to have more time and money to attend to environmental needs and amenities. Thus, while richer nations consume more goods and services, including energy, per capita, they also have generally been the most aggressive in addressing pollution and other environmental insults. This empirical relationship is known as the Environmental Kuznets Curve. However, its applicability to CO_2 emissions has been questioned,[41] and to the degree that it does exist for conventional pollutants such as sulfur dioxide, it reflects policy choices to constrain emissions.

These interactions have both short-run and long-run implications. For most nations most of the time, the combination of population growth and per capita GDP growth has more than offset forces tending to depress emissions, so emissions have increased. Overall, the most critical interaction is the one between per capita GDP growth and resource uses, especially energy, but also including cement manufacture, agricultural practices, deforestation, waste disposal, and the consumption and release of certain chemicals.

Changes in Intensity to Meet Climate Stabilization Goals

What might be required to "prevent dangerous interference with the climate system" remains debatable. The answer actually depends on the concentration of greenhouse gases in the atmosphere, not the level of emissions at a given point in time. Ultimate goals, then, are typically expressed in terms of what concentration would be required to keep global warming below a certain amount with a certain probability.[42] Models are then used to assess what emission reductions would be required to keep concentrations below the target level.

As noted earlier, the Copenhagen Accord is an agreement of the parties to the UNFCCC to begin establishing actions to follow on the Kyoto Protocol. The Copenhagen Accord[43] does not mandate specific reductions, but sets a goal of reducing global emissions "so as to hold the increase in global temperature below 2 degrees C, and take action to meet this objective consistent with science and on the basis of equity."

The United States has declared that its Copenhagen Accord target commitment is for a quantified economy-wide emissions reduction for 2020 "in the range of 17%" from 2005, "in conformity with anticipated U.S. energy and climate legislation, recognizing that the final target will be reported to the Secretariat in light of enacted legislation."[44]

U.S. Greenhouse Gas Intensity: Trends and Targets

Analyzing and projecting the values and the rates of change for the variables population, income, intensity, and emissions depend on the baseline, the time period in question, and assumptions about changes over time. For the purpose of thinking about the United States slowing and then reversing its increase in greenhouse gas emissions, the historic trends in population, income growth, and greenhouse gas intensity indicate the magnitude of the challenge:

United States: Trends in Drivers of GHG Emissions, 1990-2005 (Annual rate of change, %)						
Population		Per Capita GDP		GHG Intensity		GHG Emissions
(1.1)	+	(1.8)	+	(-1.9)	=	(1.0)

To simply stop the growth in GHG emissions, the three factors on the left side of the equation must sum to zero. Population growth is slowing: the actual rate for 2000-2006 was slightly over 0.9%, and the U.S. Census Bureau projects it to fall to +0.8% by 2050.[45] Assuming that population continues to grow at +0.9% through 2020, and that per capita GDP grows at the rate of +1.8% of 1990-2005, then GHG intensity would have to decline at the rate of -2.7% per year to stabilize emissions.

But the U.S. target for the Copenhagen Accord is to reduce GHG emissions for 2020 to 17% below 2005 emissions. Taking the CAIT emissions data for 2005 of 1,892 MMTCE (excludes land use changes and forestry and international bunkers[46]), the 2020 target would be 1,570 MMTCE. This would require an emissions reduction of -1.2% per year for 2005-2020 (actually, emissions grew through 2007, before turning down during the recession in 2008). If one assumes that 2010 GHG emissions were at the 2005 level and that the annual trends of +0.9% population and +1.8% per capita GDP continue[47] through 2020, what rate of intensity decline would be necessary to achieve the 2020 goal? The answer is, it would take a rate of intensity decline of about -4.6% per year[48] beginning in 2010, to reach the level of 1,570 in 2020.

United States: Trends in Drivers of GHG Emissions, 2010-2020, to Meet Copenhagen Accord Target (Annual rate of change, %)						
Population		Per Capita GDP		GHG Intensity		GHG Emissions
(0.9)	+	(1.8)	+	(-4.6)	=	(-1.9)

This represents a substantial, ongoing improvement in intensity, from a GHG intensity of 153 MMTCE/million$PPP in 2005, to an intensity of 86 in 2020—but perhaps not impossible, when one considers that in 2005 France's intensity level was 80.

Over the longer term, much more aggressive goals have been proposed: the U.S. target for the Copenhagen Accord appended a note suggesting a goal for 2050 of an 83% reduction from 2005 levels of GHG emissions,[49] which would limit U.S. emissions to 321 MMTCE. Assuming population and per capita GDP grow from 2010 to 2050 at the average annual rates of 0.85% and 1.8%, respectively, then given the emission rate at the cap, U.S. greenhouse gas intensity in 2050 would be about 8 MMTCE/million$PPP—an extremely low-carbon economy. Or, in terms of rate of change, intensity would have to decline between 2005 and 2050 at an average rate of about -6.6% per year.[50]

To give perspective to rates of intensity decline, consider an illustrative scenario in which, for each of the 10 years 2016-2025, two 1,000-megawatt nuclear electrical generating facilities go into service (or equivalent generating capacity based on renewables), replacing existing coal-fired facilities. Each plant would displace approximately 6 million tons of carbon per year; when all 20 coal-supplanting plants were in service in 2025, they would be displacing 120 million tons of carbon per year. All else equal, displacing this much carbon would accelerate the rate of decline in intensity for 2016-2025 by about -0.5% per year. This example, which lowers emissions and intensity only incrementally, shows that large declines in intensity would require multiple initiatives. To meet the goal of reducing economy-wide emissions to 17% of 2005 levels by 2050 implies some mix of making tremendous gains in energy efficiency, shifting to energy sources that emit virtually no CO_2, and developing the capacity to capture and sequester enormous amounts of CO_2

Global Greenhouse Gas Intensity

As has been noted, world greenhouse gas intensity has been declining, but not at a rate sufficient to prevent rising GHG emissions (numbers do not add precisely, due to rounding):

World GHG Emission Drivers, 1990-2005 (excludes land use changes and forestry and international bunker) (Annual rate of change, %)						
Population		Per Capita GDP[51]		GHG Intensity		GHG Emissions
(1.4)	+	(1.7)	+	(-1.6)	=	(1.6)

An in-depth analysis of policies and programs for reducing global greenhouse gas emissions is far beyond the scope of this report. But if greenhouse gases are to be reduced, the imperative to reduce intensity is clear. Simply put, more people at higher standards of living means more goods and services, especially energy—to cook, heat and cool homes, to manufacture goods, to transport people and goods, etc.

To decouple those increases in the numbers of consumers and their consumption from increases in greenhouse gas emitting energy uses implies policies fostering greater efficiency in using energy and/or use of non-greenhouse gas-emitting forms of energy, such as renewables or nuclear. But greater efficiency ultimately reaches limits from the laws of physics; alternative fuels run into the facts that, in most places, fossil fuel, either coal or natural gas, is the least expensive fuel for generating electricity and heat, and oil is the least expensive fuel for powering transport.

Beyond the energy sector, moreover, there are many other areas where policies may affect GHG emissions. Land use and agricultural and forestry policies can have direct implications for emissions, and could reduce intensity. The non-CO2 gases, many of which pose particularly long-term climate implications, offer cost-effective opportunities for reductions from certain industrial processes, landfills, and fuel production.

Perhaps most importantly, at the global scale, the possibility exists for identifying and exploiting least-expensive opportunities for reducing greenhouse gases, thereby increasing the efficiency with which economies use greenhouse gas-emitting technologies. This depends, however, on global instruments for accounting for and verifying such reductions. Reaching practical agreements on international mechanisms (e.g., for a carbon tax or a cap-and-trade system to obtain economic efficiencies among nations in reducing emissions) requires divergent national goals to be focused on what is, ultimately, a global issue. The global nature of climate change challenges national sovereignty. The UNFCCC, the Kyoto Protocol, the Asia-Pacific Partnership, and the Copenhagen Accord are efforts in multilateral approaches to reducing emissions, but their individual and complementary successes remain to be seen.

CONCLUSION

In the end, the interactions of the variables, population, income, and intensity of emissions **(equation 1)**, together with the inexorable force of compounding growth rates over time **(equation 2)** are inescapable conditions determining both the risks of climate change and the costs, benefits, and tradeoffs of options for responding. If climate change poses a genuine risk to the well-being of mankind, the nations of the world, individually and collectively, face two fundamental challenges: adopting and implementing policies and encouraging the development and use of technologies that emit lower levels of greenhouse gases, and maintaining a sufficiently high rate of intensity decline over the long term to ensure declining emissions.

In 1992, Congress enacted the Energy Policy Act of 1992 (EPACT, P.L. 102-486), which contained provisions to implement the United Nations Framework Convention on Climate Change (UNFCCC), which had been signed earlier in the year.[52] The UNFCCC's objective to stabilize "greenhouse gas concentrations in the atmosphere at a level that would prevent dangerous anthropogenic interference with the climate system" was echoed in EPACT, which called for a National Energy Policy Plan to "include a least-cost energy strategy ... designed to achieve [among other goals] ... the stabilization and eventual reduction in the generation of greenhouse gases.... "[53]

In ratifying the UNFCCC, the United States agreed to several principles for achieving this objective, including the following:

- "[D]eveloped country parties should take the lead in combating climate change and the adverse effects thereof."[54]
- "Parties should take precautionary measures to anticipate, prevent or minimized the causes of climate change and mitigate its adverse effects."[55]

- "Parties have a right to, and should, promote sustainable development...."Climate change policies should take "into account that economic development is essential for adopting measures to address climate change."[56]

The UNFCCC's linking of sustainable development and climate change mitigation reflects the perceived need to decouple economic development and growth from non-sustainable, greenhouse gas-emitting energy technologies.

As this report suggests—

- An expanding global population is an important driver for economic growth. As affirmed in the UNFCCC, climate change policies are to take "into full account the legitimate priority needs of developing countries for the achievement of sustained economic growth and the eradication of poverty."[57]
- Economic development may reduce population pressure in the long-term but creates increasing demand for resources that, employing current technologies, contribute to greenhouse gas emissions. Although economies become more efficient over time, those efficiencies have yet to overcome the combination of expanding population and growing economies without the intervention of governments.
- Satisfying the energy needs of dynamic economies is increasing the demand for coal and other fossil fuels for economic and other reasons. To meet the massive reductions in greenhouse gas emissions in the long term required by various stabilization scenarios would seem to require the development and deployment of commercially available technologies to shift economies substantively away from fossil fuels, and/or the large-scale capture and sequestration of the emissions of CO_2 from coal and other fossil fuels. The UNFCCC recognizes the "special difficulties of those countries, especially developing countries, whose economies are particularly dependent on fossil fuel production, use and exportation, as a consequence of action taken on limiting greenhouse gas emissions."[58]

Breaking the current dynamic of increasing populations and economic growth pushing up greenhouse emissions would depend on developing "sustainable" alternatives—both in improving the efficiency of energy use and in moving the fuel mix toward less greenhouse gas-emitting alternatives. In the

UNFCCC, developed nations committed to taking the initiative by "adopt[ing] national policies and tak[ing] corresponding measures on the mitigation of climate change ... [that] will demonstrate that developed countries are taking the lead in modifying longer-term trends in anthropogenic emissions consistent with the objective of the Convention...."[59] Such development paths are critical not only for any domestic program, but also participation by developing countries in any global greenhouse gas stabilization program may be at least partially dependent on the availability and cost of such technologies.

As stated by the UNFCCC,

> The extent to which developing country parties will effectively implement their commitments under the Convention will depend on the effective implementation by developed country parties of their commitments under the Convention related to financial resources and transfer of technology and will take fully into account that economic and social development and poverty eradication are the first and overriding priorities of the developing country parties.[60]

The Copenhagen Accord places new emphasis on technology development and its transfer among nations and represents an important component of the United States'—and other developed nations'—response to this principle. It remains to be seen whether the proposed funds are forthcoming, and how they are dispersed. Fostering technological change depends on two driving factors: exploiting new technological opportunities (technology-push) and market demand (market-pull).[61]

Currently, U.S. policy is oriented primarily to the technology-push part of the equation, with a focus on research and development (R&D). In contrast, the European Union (EU) is complementing its research and development efforts by constructing a multi-phased, increasingly more stringent market-pull for greenhouse gas-reducing technologies and approaches, including taxes and regulatory requirements overlain by the EU's Emissions Trading System.[62]

The market-pull side focuses on market interventions to create demand, which poses questions of—

- Whether, how, and to what extent to use price signals to change behaviors and to stimulate innovation of technologies that increase energy efficiency or that emit less greenhouse gases. Direct taxes on energy or on greenhouse gases could be one approach, whereas the concept of shifting taxes from incomes to consumption would be a broader one.

- Whether, how, and to what extent to use regulatory actions to change behaviors and to require technologies that increase energy efficiency or emit less greenhouse gases. A direct regulatory effort would be a renewable power standard for electricity-generating facilities, which requires some specified portion of electric power to be generated by renewables, such as water power, solar, or wind (whether nuclear power might count is an open question).Heretofore, especially in the United States, regulatory efforts curtailing greenhouse gas emissions commonly originated in response to other objectives, such as reducing health-damaging air pollutants or enhancing energy security by fostering substitutes for imported oil. In these cases, reductions in greenhouse gases were coincidental ("no regrets"); further co-reduction opportunities remain(e.g., methane from landfills). However, the objective of reducing greenhouse gases as the primary object of regulations is increasingly coming to the fore, especially in some states.[63]

The technology-push side focuses on research and development. It raises questions as to what R&D programs should be supported at what levels:

- Over the short-to mid-term, how can existing technologies be made more sustainable? How can energy (and other resources) be used more efficiently? What alternatives can be pursued?
- What are the relative federal and private roles in selecting and financing R&D of specific technologies?
- Perhaps most important for the longer run, what breakthrough research should be pursued? Over the past 100 years, a number of technological changes have occurred (e.g., in nuclear power, computing, and communications) that demonstrate the low success rate of predicting technological and societal changes far into the future. At present, at least two technological breakthrough possibilities can be discerned: fusion power, which conceivably could wean economies from fossil fuels, and sequestration,[64] which could capture and store carbon dioxide—and perhaps even remove excess from the atmosphere. Other breakthroughs are surely possible—including serendipitous discoveries that cannot be conceived of now.

If the ultimate, 2050 target for reducing greenhouse gas emissions is as aggressive as 83% below 2005 levels, as in some proposals, then fundamentally at issue is whether the risks of climate change can be addressed only by incremental "muddling through" or whether some extraordinary, aggressive effort is needed. Certainly, there are many opportunities for incremental and iterative policies to reduce greenhouse gases, to conserve energy, to find alternative energy sources, to make vehicles more energy efficient, to enhance carbon sequestration through afforestation and refined cropping practices, to deter deforestation and land use changes that increase CO2 emissions, and so on. In addition, the potential for a substantial supply of natural gas offers the United States the possibility of buying time (similar to what the U.K. did in the 1990s) to allow a combination of research and development and market forces responding to a government-imposed carbon price to providing longer term opportunities. The incremental nature of such a response provides flexibility, while a time frame of decades offers hope of unpredictable breakthroughs or the discovery that climate change is not so threatening as some fear.

Conversely, given the drivers increasing emissions, such as population growth and economic development and growth, it is hard to see how incremental changes affecting intensity will achieve the rate of intensity decline sufficient to reduce emissions to the proposed levels, even over decades.[65] From this perspective, an intense, aggressive pursuit of break-throughs—even with high costs and high risks of failure—has to be weighed against the costs and risks of failing to prevent potentially dangerous interference with the climate system.

End Notes

[1] CRS Report RL33849, *Climate Change: Science and Policy Implications*, by Jane A. Leggett, for more information.

[2] For a review of international activities, see CRS Report RL33826, *Climate Change: The Kyoto Protocol, Bali "Action Plan," and International Actions*, by Jane A. Leggett. For a review of U.S. activities, see CRS Report R40556, *Market-Based Greenhouse Gas Control: Selected Proposals in the 111th Congress*, by Jonathan L. Ramseur, Larry Parker, and Brent D. Yacobucci. .

[3] UNFCCC, Article 2, "Objectives."

[4] Charter for the Asia-Pacific Partnership on Clean Development and Climate (January 12, 2006), "Purposes," 2.1.1. For additional information, see http://www.asiapacificpartnership.org/.

[5] http://en.cop15.dk/files/pdf/copenhagen_accord.pdf

[6] See, for example, CRS Report RL32721, *Greenhouse Gas Emissions: Perspectives on the Top 20 Emitters and Developed Versus Developing Nations*, by Larry Parker and John Blodgett;

CRS Report RL32762, *Greenhouse Gases and Economic Development: An Empirical Approach to Defining Goals*, by John Blodgett and Larry Parker; *Climate Data: Insights and Observations* (World Resources Institute; prepared for the Pew Center on Global Climate Change, December 2004), http://cait.wri.org/.

[7] World Resources Institute, Climate Analysis Indicators Tool (CAIT), as described below.

[8] In particular, following the breakup of the former Soviet Union, the economies of various Eastern European and former Soviet republics contracted in the 1990s, such that their emissions declined substantially between 1990 and 2000.

[9] This database uses a variety of data sources to provide information on greenhouse gas emissions and other relevant indicators. Full documentation, along with caveats, is provided on the WRI website at http://cait.wri.org/.

[10] See http://www.census.gov/ipc/www/idb/worldpop.html.

[11] GDP-PPP is gross domestic product converted into international dollars using purchasing power parity (PPP) rates. An international dollar has the same purchasing power in the domestic currency as a U.S. dollar has in the United States. The World Bank is the source of CAIT's PPP data. (World Resources Institute, *CAIT: Indicator Framework Paper*, December 2008, p. 23.)

[12] Emissions comprise six greenhouse gases: carbon dioxide, nitrous oxide, methane, perfluorocarbons, hydrofluorocarbons, and sulfur hexafluoride. To aggregate emissions data, figures are typically given in millions of metric tons of carbon equivalents (MMTCE). Thus, global aggregate greenhouse gas emissions (excluding land use changes) for 2005 were 10,569 MMTCE, or 10.6 billion tons. In the text, unless otherwise explicitly stated, "tons" of emissions means "metric tons of carbon equivalents." To convert carbon equivalents to CO_2 equivalents, multiply by44/12.

[13] The year 2005 is the most recent year for which CAIT has data for all six greenhouse gases. Note that analyses based on 1990-2005 data are affected by the collapse of the former USSR and do not take into account the most recent rapid increases in energy use and emissions for India and China.

[14] The rate of population growth has declined in many countries in recent decades, partly as a result of deliberate policies (e.g., birth control programs and, in a few countries, such as China, limits on family size); and partly as a result of education, higher standards of living, and cultural changes. Global population growth is expected to continue at least to mid-century, with projections suggesting a global population of around 9 billion in 2050.

[15] In principle, these figures could be calculated by adding the three left-hand data columns; in fact, a number of rows do not add; this may be due to rounding or, where discrepancies are large, from shortcomings in the underlying reported data. Nevertheless, the figures are consistent with the generalizations about trends.

[16] For Iran and Spain, population and intensity both increased.

[17] For Iran, GHG emissions rose because population and GDP growth had no offset at all from intensity, which worsened.

[18] Emissions could also be stabilized by declines in population or GDP. However, because U.S. policymakers are unlikely to promote population reduction or GDP contraction, analysis of these options seems unwarranted. In some countries (e.g., China), deliberate efforts to constrain population do occur.

[19] That is, annual population growth (1.5%) + per capita GDP growth (1.4%) + intensity change (-2.9% [rather than the actual -2.0%]) = 0 emissions growth.

[20] That is, annual population growth (1.2%) + per capita GDP growth (2.0%) + intensity change (-3.2% [rather than the actual -1.9%]) = 0 emissions growth.

[21] Rates of change were not calculated if both the 1990 and 2005 emission levels were below 5 million tons. At low levels, even small changes can yield notable rates of change—for example, if emissions went from 2 to 4 million tons between 1990 and 2005, the rate of change would be 3.8% per year, but the actual emissions are too small to meaningfully affect overall totals.

[22] Including CO_2 from cement manufacture, N_2O from Adipic and Nitric Acid production, N_2O and CH_4 from other industrial processes, plus HFCs, PFCs, and SF_6.

[23] See Timothy Herzog et al., *Target: Intensity, An Analysis of Greenhouse Gas Intensity Targets* (Washington, DC: World Resources Institute, November 2006), pp. 3-9; and Frank Princiotta, "Global Climate Change and the Mitigation Challenge, *Journal of the Air & Waste Management Association* (October 2009), Vol. 59, pp.1194-1211.

[24] As given, the emissions data are taken from CAIT tables, but in principle could be calculated by multiplying the intensity (column 4) times GDP; because of inconsistent data, the calculations in some cases diverge from the reported emissions, though the general magnitudes and the relative positions of nations are right.

[25] In principle, the sum of the first two data columns should equal the third data column, and the fifth column should be the sum of the third and fourth columns; however, because of data inconsistencies, the calculated numbers may not exactly correspond to the CAIT reported numbers. Nevertheless, the general relationships hold.

[26] Environmental Protection Agency, *Human-Related Sources and Sinks of Carbon Dioxide* http://www.epa.gov/climatechange/emissions/co2_human.html

[27] Energy Information Administration, *International Energy Statistics.* http://tonto.eia.doe.gov/cfapps/ipdbproject/IEDIndex3.cfm?tid=2&pid=2&aid=12 Totals add to slightly more than 100% because offsetting pump-storage generation is not included.

[28] International Energy Agency, *Electricity Information 2002* (OECD, 2002), p. II.285.

[29] Climate analysis Indicators Tool (CAIT), version 7.0 (Washington, DC: World Resources Institute, 2010).

[30] International Energy Agency, *Electricity Information 2002* (OECD, 2002), p. II.683.

[31] If gas combined-cycle technology is considered, the ratio could be 0.4 or 0.5 to 1.

[32] Climate Analysis Indicators Tool (CAIT), version 7.0. (Washington, DC: World Resources Institute, 2010).

[33] Climate Analysis Indicators Tool (CAIT), version 7.0. (Washington, DC: World Resources Institute, 2010).

[34] Feng An, et al., *Passenger Vehicle Greenhouse Gas and Fuel Economy Standards: A Global Update*, International Council on Clean Transportation (July 2007), p. 8.

[35] For example, the London "congestion tax" is intended to shift commuters out of passenger cars and onto public transit.

[36] See CRS Report R40155, *Selected Issues Related to an Expansion of the Renewable Fuel Standard (RFS)*, by Randy Schnepf and Brent D. Yacobucci.

[37] This is not a simple cause and effect, but reflects evolutionary changes in areas such as education, cultural expectations, women's rights, access to birth control, and health care—all of which may be affected by social policy.

[38] For example, for the United States, the U.S. Census Bureau projected that in 2010, U.S. population growth would be 1%, with immigration accounting for just under half the increase (about 45%). For 2050, the projection is a growth rate of about 0.8%, with immigration accounting for more than half the increase (about 60%). http://www.census.gov/population/www/projections/summarytables.html

[39] See http://www.un.org/News/Press/docs//2007/pop952.doc.htm.

[40] CRS Report RL32721, *Greenhouse Gas Emissions: Perspectives on the Top 20 Emitters and Developed Versus Developing Nations*, by Larry Parker and John Blodgett.

[41] The entire March issue of the *Journal of Environment & Development*, vol. 14 (2005) is devoted to this topic; see especially Joseph E. Aldy, "An Environmental Kuznets Curve Analysis of U.S. State-Level Carbon Dioxide Emissions," pp. 48-72. Also, William R. Moomaw and Gregory C. Unruh, "Are Environmental Kuznets Curves Misleading Us? The Case of CO2 Emissions," in *Environment & Development Economics* (Cambridge University Press, 1997), pp. 451-463.

[42] See, for example, M.G.J. den Elzen and M. Meishausen, "Meeting the EU 2°C climate target: global and regional emission implications," Report 728001031/2005, Netherlands Environmental Assessment Agency.

[43] http://en.cop15.dk/files/pdf/copenhagen_accord.pdf

[44] http://unfccc.int/home/items/5264.php. The U.S. commitment adds the following note: "The pathway set forth in pending legislation [i.e., H.R. 2454, S. 1733] would entail a 30% reduction in 2025 and a 42% reduction in 2030, in line with the goal to reduce emissions 83% by 2050."

[45] http://www.census.gov/population/www/projections/summarytables.html

[46] Accounting for these exclusions changes the emissions only to 1,896 MMTCE, or +0.2%—irrelevant in terms of likely errors in reported emissions.

[47] It is important to recognize that we are looking at trends over an extended time; these assumed average trends blur short term variations (e.g., higher rates of intensity decline in 2004-2005, or possible variations from the recession that started in 2008). *The Economic Report of the President, 2010*, forecasts growth rates through 2020 (see Table 2-3, p. 75), that rise to over 4% for 2011-2013, then decline to 2.5% for 2019-2020—or in terms of the per capita datapresented in this report, subtracting population growth gives rates of approximately 2.5% annual per capita growth for2011-2013, declining to 1.7% for 2019-2020.

[48] Achieving this could include, besides direct reductions in emissions, offsets from reductions made and paid for in other countries, as well as reductions from land use changes and sequestration.

[49] http://unfccc.int/home/items/5264.php The U.S. commitment adds the following note: "The pathway set forth in pending legislation would entail a 30% reduction in 2025 and a 42% reduction in 2030, in line with the goal to reduce emissions 83% by 2050."

[50] Achieving this could include, besides direct reductions in emissions, offsets from reductions made and paid for in other countries, as well as reductions from land use changes and sequestration.

[51] Economic contractions of several newly independent nations following the breakup of the former Soviet Union depressed global GDP, so this rate will likely rise in subsequent decades.

[52] The United States signed the UNFCCC on June 12, 1992, and ratified it on October 15, 1992 The UNFCCC entered into force on March 21, 1994.

[53] Section 1602(a)

[54] UNFCCC, article 3.

[55] Ibid.

[56] Ibid.

[57] UNFCCC, Preamble.

[58] UNFCCC, Preamble.

[59] UNFCCC, article 4(2)(a).

[60] UNFCCC, article 4(7).

[61] L. Clarke, J. Weyant, and A. Birky, "On the Sources of Technological Change: Assessing the Evidence," *Energy Economics*, vol. 28 (2006), pp. 579-595.

[62] See CRS Report R41049, *Climate Change and the EU Emissions Trading Scheme (ETS): Looking to 2020*, by Larry Parker.

[63] See CRS Report RL33812, *Climate Change: Action by States to Address Greenhouse Gas Emissions*, by Jonathan L. Ramseur.

[64] See CRS Report RL33801, *Carbon Capture and Sequestration (CCS)*, by Peter Folger.

[65] However, some pollution control efforts have had dramatic successes: lead has been essentially eliminated as an air pollutant; regulated auto emissions have been reduced by over 90% from unregulated levels; between 1990 and 2005,sulfur dioxide emissions from acid rain program sources dropped by about 35%; and electricity generated rose about30%.

In: Greenhouse Gas Emissions
Editors: M. Knox and J. D. Mark
ISBN 978-1-62081-287-7

Chapter 2

CLIMATE CHANGE: THE QUALITY, COMPARABILITY, AND REVIEW OF EMISSIONS INVENTORIES VARY BETWEEN DEVELOPED AND DEVELOPING NATIONS*

United States Government Accountability Office

WHY GAO DID THIS STUDY

Nations that are parties to the United Nations Framework Convention on Climate Change periodically submit inventories estimating their greenhouse gas emissions. The Convention Secretariat runs a review process to evaluate inventories from 41 "Annex I" nations, which are mostly economically developed nations. The 153 "non-Annex I" nations are generally less economically developed and have less stringent inventory reporting guidelines. The Department of State (State) represents the United States in international climate change negotiations. GAO was asked to report on (1) what is known about the comparability and quality of inventories and barriers, if any, to improvement; (2) what is known about the strengths and limits of the

* This is an edited, reformatted and augmented version of the Highlights of GAO-10-818, a report to congressional requesters, dated July 2010.

inventory review process; and (3) views of experts on implications for current and future international agreements to reduce emissions. GAO analyzed inventory reviews and inventories from the seven highest-emitting Annex I nations and seven of the highest emitting non-Annex I nations. GAO also selected and interviewed experts.

WHAT GAO RECOMMENDS

GAO recommends that the Secretary of State work with other parties to the Convention to (1) continue encouraging non-Annex I parties to improve their inventories and (2) strengthen the inventory review process's quality assurance framework. State agreed with GAO's findings and recommendations.

WHAT GAO FOUND

Recent reviews by expert teams convened by the Secretariat found that the 2009 inventories from the selected Annex I nations—Australia, Canada, Germany, Japan, Russia, the United Kingdom, and the United States—were generally comparable and of high quality. For selected non-Annex I nations—Brazil, China, India, Indonesia, Malaysia, Mexico, and South Korea—GAO found most inventories were dated and of lower comparability and quality. Experts GAO interviewed said data availability, scientific uncertainties, limited incentives, and different guidelines for non-Annex I nations were barriers to improving their inventories. The lack of comparable, high quality inventories from non-Annex I nations is important because they are the largest and fastest growing source of emissions, as shown in the figure, and information about their emissions is important to efforts to address climate change. There are no inventory reviews for non-Annex I nations.

Experts said the inventory review process has notable strengths for Annex I nations as well as some limitations. The review process, which aims to ensure nations have accurate information on inventories, is rigorous, involves well-qualified reviewers, and provides feedback to improve inventories, according to experts. Among the limitations experts identified is a lack of independent verification of estimates due to the limited availability of independent statistics against which to compare inventories' data. Also, GAO found that the review process's quality assurance framework does not independently assess concerns about a limited supply of reviewers and inconsistent reviews, which could pose challenges in the future.

Experts said Annex I nations' inventories and the inventory review process are generally sufficient for monitoring compliance with current agreements to reduce emissions. For non-Annex I nations, however, experts said the current system may be insufficient for monitoring compliance with future agreements, which may require more reporting. As part of ongoing negotiations to develop a new climate change agreement, State has emphasized the need for better information on emissions from high-emitting non-Annex I nations. While improving the inventory system is important to negotiations, some experts said disagreements about emissions limits for developed and developing nations pose a greater challenge.

ABBREVIATIONS

EPA	Environmental Protection Agency
IEA	International Energy Agency
IPCC	Intergovernmental Panel on Climate Change

July 30, 2010

The Honorable Joe Barton
Ranking Member
Committee on Energy and Commerce
House of Representatives

The Honorable Michael C. Burgess
Ranking Member
Subcommittee on Oversight and Investigations
Committee on Energy and Commerce
House of Representatives

The Honorable Greg Walden
House of Representatives

High-quality information on greenhouse gas emissions is critical to domestic and international efforts to address climate change. Elevated concentrations of greenhouse gases in the atmosphere could alter the climate and adversely affect agriculture, infrastructure, ecosystems, and human health. To address these risks, Congress is considering proposals to limit greenhouse gas emissions, and the United States is participating in international negotiations to develop an international response to climate change.

In 1992, the United States and most nations of the world negotiated the United Nations Framework Convention on Climate Change (the Convention) to stabilize atmospheric concentrations of carbon dioxide and five other greenhouse gases.[1] As a step toward that goal, under the 1997 Kyoto Protocol to the Convention, 37 industrialized nations and the European Community agreed to binding emissions targets for 2008 through 2012.[2] Many nations with significant greenhouse gas emissions, including China, India, and the United States, have not committed to such binding targets in the Kyoto Protocol or subsequently.[3] However, all nations that are parties to the Convention agreed, among other things, to periodically provide inventories detailing their man-made emissions and removals of greenhouse gases.[4] These inventories, and processes for their review, play an important role in ongoing negotiations for a post-2012 agreement to extend or succeed the Kyoto Protocol. In December 2009, these negotiations resulted in the Copenhagen Accord, a nonbinding political agreement in which, among other things, certain nations announced various actions to reduce emissions and developing nations agreed to submit more frequent reports on their emissions.

Forty-one of the Convention's parties—most of them economically developed nations—are listed in Annex I of the Convention and submit annual inventories of their greenhouse gas emissions.[5] Developing an inventory involves collecting data on activities across all sectors of a nation's economy that influence emissions and using numerous methods to estimate associated emissions. Annex I nations' inventories undergo a review process coordinated by the Convention's Secretariat, with reviews by teams of international experts. The review teams evaluate consistency with inventory guidelines agreed to by all parties, including technical methods developed by the Intergovernmental Panel on Climate Change (IPCC).[6] An additional 153 parties to the Convention are not listed in Annex I and are known as non-Annex I nations. They are generally less economically developed, though strong economic growth since 1992 in some nations means their per-capita income now surpasses that of some Annex I nations. They are encouraged to use the same technical methods as Annex I nations in developing their inventories. However, the parties agreed that inventories from non-Annex I nations do not need to be submitted as often, include estimates for as many gases, or undergo the same reviews as inventories from Annex I nations. In 2003, we reported that select Annex I nations varied in their adherence to reporting standards and that there was little information on inventories from select non-Annex I nations.[7]

The role of certain non-Annex I nations has become central to negotiations on a post-2012 agreement because their emissions have increased substantially since 1997 and are expected to continue to grow. China, a non-Annex I nation, recently overtook the United States as the world's largest emitter, according to some estimates.[8] According to Energy Information Administration projections, non-Annex I nations may contribute nearly all of the growth of global fossil-fuel related emissions through 2030.[9] Because of this expected growth, emissions reductions will be needed from high-emitting nations, including non-Annex I nations, to stabilize the concentration of greenhouse gases in the atmosphere. Recognizing this, the United States' position in international negotiations has been that a new climate change agreement should extend binding actions and strong requirements for monitoring, reporting, and verification of emissions to large non-Annex I nations. In negotiations, many non-Annex I nations have taken the position that Annex I nations are obliged to begin to cut their emissions more deeply because they have emitted a large share of the greenhouse gases that currently drive climate change.

In the United States, the Environmental Protection Agency (EPA) prepares the annual inventory with contributions from other agencies, including the Departments of Agriculture and Energy.[10] The Department of State (State) represents the United States in international negotiations to develop a post-2012 agreement to address climate change, and participates in the assessment and review of whether the Convention is being effectively implemented, including the inventory review process. State also officially submits the U.S. inventory to the Convention's Secretariat.

In response to your request, this report addresses the following questions: (1) what is known about the comparability and quality of inventories submitted by developed and developing nations and barriers, if any, to improving comparability and quality; (2) what is known about the strengths and limits of the Convention's inventory review process; and (3) what are the views of experts of the implications of the state of the inventory system for establishing or monitoring compliance with international agreements to reduce greenhouse gas emissions.

To provide information on the comparability and quality of inventories, we summarized the results of the most recent inventory reviews of the seven largest emitting Annex I nations—Australia, Canada, Germany, Japan, Russia, the United Kingdom, and the United States. For non-Annex I nations, we assessed whether inventories from seven of the largest greenhouse gas emitting nations—Brazil, China, India, Indonesia, Malaysia, Mexico, and South Korea—included estimates for all major greenhouse gases, for all economic sectors, and for various years, among other factors. In 2005, the latest year for which global estimates are available, these 14 nations represented about two-thirds of the greenhouse gas emissions that were not related to land use and forestry.[11] We did not independently assess the comparability or quality of inventories from Annex I nations or of emissions estimates from non-Annex I nations. Though we identified some limitations with the inventory review process, we believe that the reviews provide reasonable assessments of the comparability and quality of inventories from the selected Annex I nations. Our findings are not generalizable to other nations because the selected nations are not necessarily representative. To address all three objectives, we also summarized findings in the literature and the results of semi-structured interviews with 15 experts on national emissions inventories, the Convention's inventory review process, and international agreements. We selected these experts based on recommendations from U.S. and international government officials and researchers, the relevance and extent of their publications, and their involvement in the Convention's inventory review process and related activities. We ensured that our group of experts covered key perspectives, including the views of Annex I and non-Annex I nations and the views of experts not directly involved in preparing or reviewing inventories. Experts included agency and international officials, researchers, and members of inventory review teams. Not all of the experts provided their views on all issues. We identify the number of experts providing views where relevant. Appendix I provides additional information about our scope and methodology, and appendix II lists the experts we interviewed.

We conducted this performance audit from September 2009 to July 2010 in accordance with generally accepted government auditing standards. Those standards require that we plan and perform the audit to obtain sufficient, appropriate evidence to provide a reasonable basis for our findings and conclusions based on our audit objectives. We believe that the evidence obtained provides a reasonable basis for our findings and conclusions based on our audit objectives.

BACKGROUND

Greenhouse gases can affect the climate by trapping energy from the sun that would otherwise escape the earth's atmosphere. Various human and natural activities emit greenhouse gases, with the production and burning of fossil fuels for energy contributing around two-thirds of man-made global emissions in 2005 (see figure 1). The remaining third includes emissions from industrial processes, such as steel production and semiconductor manufacturing; agriculture, including emissions from the application of fertilizers and from ruminant farm animals; land use, such as deforestation and afforestation; and waste, such as methane emitted from landfills. Carbon dioxide is the most important of the greenhouse gases affected by human activity, accounting for about three-quarters of global emissions in 2005, the most recent year for which data were available.[12]

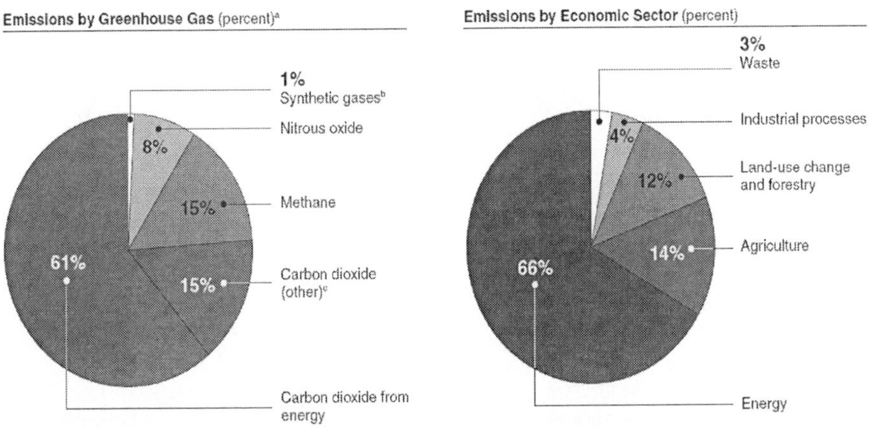

Source: GAO analysis of World Resources Institute data.

[a] Emissions are weighted by the 100-year global warming potential for each gas.

[b] Synthetic gases include hydrofluorocarbons, perfluorocarbons, and sulfur hexafluoride.

[c] Other includes carbon dioxide emissions from industrial processes, agriculture, land-use change and forestry, and waste.

Note: Figure includes emissions and removals of greenhouse gases.

Figure 1. Global Greenhouse Gas Emissions by Gas and by Economic Sector, 2005.

The 14 nations in our study differ greatly in the quantity of their greenhouse gas emissions, the sources of those emissions, and their per- capita incomes. Emissions in 2005 ranged from about 7 billion metric tons of carbon dioxide equivalent in China and 6 billion metric tons in the United States, to about 300 million metric tons in Malaysia. The contribution of various sectors to national emissions also differed across nations, with emissions from energy and industrial processes accounting for more than 70 percent of emissions in most industrialized nations and 20 percent or less of emissions in Indonesia and Brazil (see figure 2).

The Convention established a Secretariat that, among other things, supports negotiations, coordinates technical reviews of reports and inventories, and compiles greenhouse gas inventory data submitted by nations. The Secretariat has about 400 staff, located in Bonn, Germany, and its efforts related to national inventories are funded by contributions from the parties.13 For the Secretariat's core budget, parties provided $52 million for the 2008-2009 budget cycle, of which the United States contributed $9.5 million ($3.76 million in 2008 and $5.75 million in 2009), excluding fees.

-

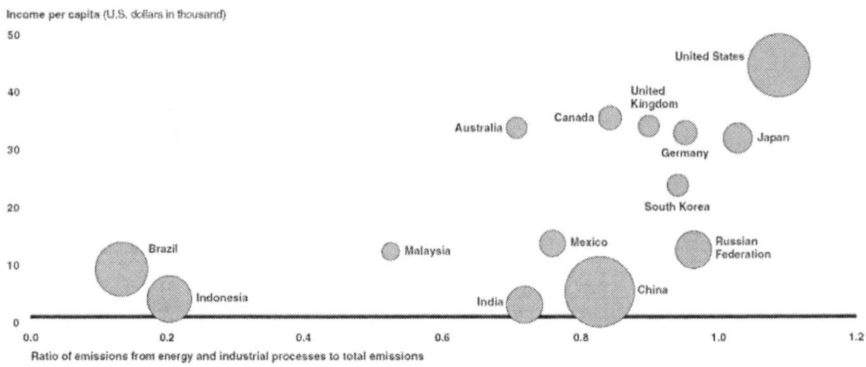

Source: GAO analysis of World Resources Institute (non-Annex I nations), Convention (Annex I nations), and World Bank data.

Notes: Size of circle represents total emissions from each nation. Since total emissions include emissions minus removals of greenhouse gases, energy and industrial process emissions are greater than total for some nations. Per capita income figures are converted from local currencies using purchasing power parities, which take into account differences in the relative prices of goods and services across nations.

Figure 2. Size and Structure of Greenhouse Gas Emissions and Per-Capita Income for 14 Selected Nations, 2005.

The Convention requires parties to periodically report to the Secretariat on their emissions of greenhouse gases resulting from human activities.14 parties do not generally measure their emissions, because doing so is not generally feasible or cost effective, and instead estimate their emissions. To help parties develop estimates, the IPCC developed detailed guidelines—which have evolved over time—describing how to estimate emissions. The general approach is to use statistics on activities, known as activity data, and estimates of the rate of emissions per unit of activity, called emissions factors.[15] For example, to estimate emissions from passenger cars, the inventory preparers could multiply the number of gallons of gasoline consumed by all cars by the estimated quantity of emissions per gallon. The IPCC guidelines allow nations to use various methods depending on their data and expertise. In some cases, with adequate data, estimates of emissions can be as accurate as direct measurements, for example for carbon dioxide emissions from the combustion of fossil fuels which contribute the largest portion of emissions for many nations.[16]

The parties agreed to the following five principles for inventories from Annex I nations:

- *Transparent.* Assumptions and methodologies should be clearly explained to facilitate replication and assessment of the inventory.
- *Consistent.* All elements should be internally consistent with inventories of other years. Inventories are considered consistent if a Party uses the same methodologies and data sets across all years.
- *Comparable.* Estimates should be comparable among parties and use accepted methodologies and formats, including allocating emissions to the six economic sectors defined by IPCC—energy, industrial processes, solvent and other product use, agriculture, land-use change and forestry, and waste.
- *Complete.* Inventories should cover all sources and sinks and all gases included in the guidelines.
- *Accurate.* Estimates should not systematically over- or underestimate true emissions as far as can be judged and should reduce uncertainties as far as practical.[17]

Annex I nations are to submit inventories annually consisting of two components—inventory data in a common reporting format and a national inventory report—both of which are publicly available on a Web site maintained by the Secretariat.[18] The common reporting format calls for emissions estimates and the underlying activity data and emissions factors for each of six sectors—energy, industrial processes, solvent and other product use, agriculture, land-use change and forestry, and waste. It also calls for data on the major sources that contribute to emissions in each sector. The inventory data are to reflect a nation's most recent reporting year as well as all previous years back to the base year, generally 1990.[19] The 2010 reporting format called for nearly 150,000 items of inventory data and other information from 1990 through 2008. The common format and underlying detail facilitate comparisons across nations and make it easier to review the data by, for example, enabling automated checks to ensure emissions were properly calculated and to flag inconsistencies in data reported over time.

The national inventory report should explain the development of the estimates and data in the common reporting format and should enable reviewers to understand and evaluate the inventory. The report should include, among other things, descriptions of the methods used to calculate emissions estimates, the rationale for selecting the methods used, and information about the complexity of methods and the resulting precision of the estimates; information on quality assurance procedures used; discussion of any recalculations affecting previously submitted inventory data; and information on improvements planned for future inventories.

The Secretariat coordinates an inventory review process that, among other things, assesses the consistency of inventories from Annex I nations with reporting guidelines. The purposes of this process are to ensure that parties are provided with (1) objective, consistent, transparent, thorough, and comprehensive assessments of the inventories; (2) adequate and reliable information on inventories from Annex I parties; (3) assurance that inventories are consistent with IPCC reporting guidelines; and (4) assistance to improve the quality of inventories.

In supporting the inventory review process, the Secretariat provides scientific and technical guidance on inventory issues and coordinates implementation of Convention guidelines. Inventory reviews are supervised by the head of the reporting, data, and analysis program within the Secretariat. By June each year, the Secretariat checks each inventory for completeness and format, called an initial check, and conducts a preliminary assessment before submitting it to an inventory review team for examination. The Secretariat assembles inventory review teams composed of scientists and other experts from around the world to review inventories from all Annex I parties according to the Convention's review guidelines. The inventory review teams assess inventories in September by reviewing activity data, emissions factors, methodologies, and other elements of an inventory to determine if a nation has employed appropriate standards, methodologies, and assumptions to compute its emission estimates. From February through March, the inventory review teams develop inventory review reports outlining their findings.

In accordance with the Convention's principle of common, but differentiated responsibilities, non-Annex I nations' inventories' format and frequency differ from those for Annex I nations. The reporting guidelines, which have evolved over time, encourage non-Annex I nations to use the IPCC methodological guidelines in developing their inventories, but do not specify that they must be used. While they submit inventories to the Secretariat, non-Annex I nations' inventories are not stand-alone documents. Rather, a non-Annex I nation's inventory is a component of its national communication, a report that discusses steps the nation is taking or plans to take to implement the Convention. Non-Annex I nations do not have to use the common reporting format or submit a national inventory report. Moreover, they do not submit an inventory each year, but instead the parties to the Convention determine the frequency of their submissions. Parties have not agreed on a regular frequency for non-Annex I nations to submit their inventories.

Inventories from Seven Annex I Nations Were of Higher Comparability and Quality than Those from Seven Non-Annex I Nations Because of Several Barriers

According to expert inventory review teams, the 2009 greenhouse gas inventories of seven Annex I nations were generally comparable and of high quality, although some of their emissions estimates have substantial uncertainty. In contrast, we found that the most recent inventories from seven non-Annex I nations, although they met reporting guidelines, were of lower quality and generally not comparable. Finally, experts identified several barriers to improving inventory comparability and quality.

Recent Reviews Found That Selected Annex I Nations' Inventories Were Comparable and of High Quality, but Some Estimates Have Substantial Uncertainty

All of the inventories submitted in 2009 by the seven selected Annex I nations were generally comparable and of high quality, according to the most recent inventory reviews conducted by expert review teams under the Convention. The reviews found that six of the seven nations—Australia, Canada, Japan, Russia, the United Kingdom, and the United States—used appropriate methodologies and data, employed reasonable assumptions, and did not systematically either over- or underestimate emissions in their 2009 inventories (covering data from 1990 through 2007). The one exception to this was Germany's 2009 inventory, which the review team said did not follow guidelines for its agricultural emissions, in part because of its attempt to use newer methods. The change significantly reduced estimated emissions from agriculture, though the sector is a relatively small contributor to Germany's total emissions. One inventory reviewer familiar with Germany's 2009 inventory said its overall quality was fairly good. In addition, Germany appears to have addressed the issue of its agricultural emissions in its 2010 inventory submission by returning to its previous methods, which had the effect of increasing its estimates of emissions from agriculture.[20] Experts said that the seven selected inventories were generally comparable, which means they generally used agreed-upon formats and methods. In addition, nine experts we interviewed said they were of high quality and did not have major flaws.

These findings show significant improvement in the seven nations' inventories since our 2003 report. For example, we reported in 2003 that both Germany's 2001 submission (covering data through 1999) and Japan's 2000 submission (covering data through 1998) lacked a national inventory report, a critical element that explains the data and methods used to estimate emissions. Nearly all Annex I nations—including Germany and Japan—now routinely submit this report. In addition, the review team found Russia's 2009 inventory showed major improvements. For example, Russia included a full uncertainty analysis for the first time and improved its quality assurance and quality control plan. Since our 2003 report, these 7 selected nations, and 34 other Annex I parties, have submitted about seven inventories, which were generally on time and more comprehensive than previous inventories (see figure 3).

The inventory review reports noted several potential problems that, while relatively minor, could affect the quality of emissions estimates. For example, the review of the 2009 U.S. inventory noted that assumptions about the carbon content of coal are outdated because they are based on data collected between 1973 and 1989. The effect on emissions estimates is not clear, but the carbon content of the coal burned as fuel may change over time, according to the inventory review report. Any such change would affect emissions, since coal is the fuel for about half of all U.S. electricity generation.[21] The U.S. inventory also used a value from a 1996 agricultural waste management handbook to estimate nitrous oxide emitted from livestock manure. The inventory review noted that livestock productivity, especially for dairy cows, has increased greatly since 1996, which would also increase each animal's output of nitrous oxide emissions. Using the IPCC's methodology for calculating emissions from excreted nitrogen, we estimated that this would lead to an underestimate of roughly 4.7 percent of total nitrous oxide emissions and 0.2 percent of total greenhouse gas emissions.[22] Finally, the review of Russia's 2009 inventory noted that it did not include carbon dioxide emissions from organic forest soils, which the inventory review report said could be significant. The inventory reviews and one expert we interviewed attributed many of the potential underestimations to a lack of data or an adequate IPCC-approved methodology and said that nations were generally working to address the issues.

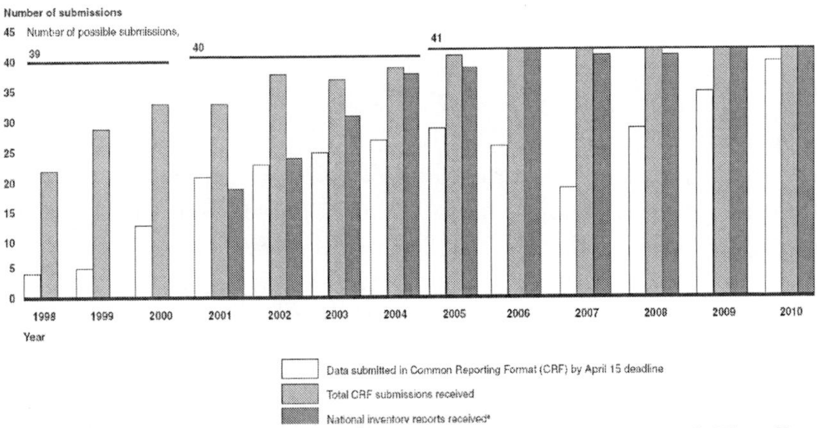

Notes: In 2007, parties submitted initial reports under the Kyoto Protocol. The effort to compile and review these reports may have contributed to the decline in on-time inventory submissions for surrounding years.

[a] Reporting guidelines called for the preparation of national inventory reports beginning in 2000, but they were not collected prior to 2001.

Figure 3. Inventory Submissions from Annex I parties, 1998-2010.

Table 1. Estimates of Uncertainty Reported in Seven Annex I Nations' 2009 Inventories for Emissions in 2007

Nation	Lower bound (percent)	Upper bound (percent)
Australia	-2%	+2%
Canada [a]	-3	+6
Germany	-10	+10
Japan	-1	+1
Russia	-40	+40
United Kingdom	-13	+13
United States	-3	+7

Source: National Inventory Reports of seven selected nations; inventory review report for Russia.

[a] Canada's uncertainty analysis does not include uncertainty from the land-use, land-use change, and forestry sector.

Even though the review teams found these seven inventories generally comparable and of high quality, the nations reported substantial uncertainty in many of the emissions estimates in their inventories. The term "uncertainty" denotes a description of the range of values that could be reasonably attributed to a quantity.[23] All of the Annex I nations' inventories we reviewed contained quantitative estimates of uncertainty. As shown in table 1, six of the seven nations reported uncertainties for their overall estimates between plus or minus 1 and 13 percent, and Russia reported overall uncertainty of about plus or minus 40 percent. That equates to an uncertainty of 800 million metric tons of carbon dioxide equivalent, slightly more than Canada's total emissions in 2007. Russia's relatively large uncertainty estimate could stem from several factors, such as less precise national statistics. In addition, Russia generally used aggregated national data rather than data that account for variation within the nation. This would increase uncertainty because aggregated data do not account for important differences that affect emissions, such as different types of technology used in the energy sector. Japan and Australia reported very low uncertainty in 2009. The inventory review report noted that Japan's estimate was lower than estimates from other nations, but neither the report nor Japan's inventory provides a full explanation. The review team for Australia said that its uncertainty ranges were generally consistent with typical uncertainty ranges reported for its sectors. Despite high levels of uncertainty in some instances, the inventory review teams found the seven inventories to be generally of high quality because the teams judge quality based on consistency with guidelines rather than strictly on the precision of the estimates.

Table 2. Greenhouse Gas Emissions and Uncertainty for Selected Categories from the 2009 U.S. Inventory for Emissions in 2007

Emissions category	Emissions estimate (in millions of metric tons of carbon dioxide equivalent)	Uncertainty of the estimate (percent)
Carbon dioxide from combustion of fossil fuels	5,736	-2% to +5%
Carbon dioxide from land use, land-use change, and forestry	-1,063 (removal of carbon dioxide)	-15 to +18
Nitrous oxide from agricultural soil management	208	-22 to +53
Methane from fermentation from digestion by livestock [a]	139	-11 to +18
Methane from landfills	133	-39 to +33
Methane from natural gas systems	105	-24 to +43

Source: GAO analysis of 2009 U.S. inventory.

[a] The technical term for this is enteric fermentation.

The uncertainty of emissions estimates also varies among the different sectors of a nation's economy. For example, uncertainty is relatively low for estimates of carbon dioxide emissions from the combustion of fossil fuels because the data on fuel use are generally accurate and the process that generates emissions is well understood. Uncertainty is much higher for certain categories within agriculture and land-use. For example, some nations report that the uncertainty in their estimates of nitrous oxide emissions from agricultural soils is greater than 100 percent, in some cases much greater. According to a March 2010 report by a National Research Council committee, this results from scientific uncertainty in emission factors.[24] Table 2 shows the contribution of the most important sources of uncertainty in the U.S. inventory. The sources of uncertainty in the other six Annex I nations' inventories follow a broadly similar pattern: the largest sources of uncertainty are either large sources of emissions—such as fossil fuel combustion and land use—or small but highly uncertain categories—such as agricultural soils.

Shortcomings in inventory reporting guidelines may decrease the quality and comparability of emissions estimates for land use, according to two experts we interviewed. For example, the guidelines state that nations should report all emissions from "managed forests," but they have broad latitude in assigning forested land to this category. This choice may have a major effect on emissions; one expert said that it would be possible for some nations with large forested areas, such as Brazil, to offset all their emissions from deforestation by designating large areas of protected forest as managed and taking credit for all of the carbon dioxide absorbed by those forests. To address this potential inconsistency, the National Research Council committee report recommended taking inventory of all land-based emissions and sinks for all lands, not just man-made emissions on managed lands. Others said that designating land as managed forest remains the most practical way to estimate man-made emissions and removals because other methods are not well developed.

Inventories from Selected Non-Annex I Nations Met Guidelines but Were of Lower Quality and Generally Not Comparable

Inventories from the non-Annex I nations we reviewed met the Convention's relevant reporting guidelines. All of the seven non-Annex I nations we reviewed—Brazil, China, India, Indonesia, Malaysia, Mexico, and South Korea—had submitted their first inventories. In addition, Mexico submitted its second, third, and fourth inventories, and South Korea submitted its second.[25] Secretariat officials said the other selected nations could submit their second inventories, as part of their national communications, over the next few years. The reporting guidelines call for non-Annex I nations to estimate emissions for 1990 or 1994 in their first submission, and for 2000 in their second submissions, and to include estimates for carbon dioxide, methane, and nitrous oxide in all submissions. We found that all selected non-Annex I nations reported for relevant years and these three gases, but we did not assess whether nations used appropriate methodologies and assumptions to develop these estimates.

However, the seven inventories were generally not comparable and were of lower quality than inventories from Annex I nations in four ways:

1. *Inventories from select non-Annex I nations were outdated.* The most recent inventories from selected Annex I nations estimate emissions for 1990-2008. However, except for Mexico and South Korea, the most recently submitted inventories from selected non-Annex I nations are for emissions for 1994.[26] (See figure 4.)
2. *Some selected non-Annex I nations' inventories do not estimate emissions of all gases.* As shown in figure 4, inventories from China, India, Indonesia, and Malaysia did not include estimates of the emissions of synthetic gases. Independent estimates show that while

synthetic gases were only 1 percent of global emissions in 2005, the emissions of synthetic gases increased by 125 percent between 1990 and 2005. Their emissions have also grown substantially in some non-Annex I nations, such as China, which had the largest absolute increase in synthetic gas emissions among all non-Annex I nations between 1990 and 2005, according to information from the International Energy Agency (IEA).[27]

3. *Select non-Annex I nations' inventories, to varying degrees, lacked critical elements.* We assessed inventories for several elements that, according to reporting guidelines, can improve the quality and transparency of inventories. First, only Brazil and Mexico provided a quantitative analysis of the uncertainty of their estimates.[28] Second, we found that all inventories lacked adequate documentation of methodologies, emission factors, and assumptions and that most lacked descriptions of quality assurance and quality control measures. Third, none of the select nations reported in a comparable format, instead using different formats and levels of aggregation.[29] For example, China estimated some methane emissions from various agricultural subsectors but grouped some of these estimates into only one category. In contrast, South Korea estimated these same emissions but reported them in separate categories. Overall, the lack of documentation and of a common reporting format limited our ability to identify and compare estimates across nations. Finally, only Mexico included an analysis of its key categories of emissions.[30]
4. *National statistics from some select non-Annex I nations are less reliable.* According to three experts we interviewed and literature, some non-Annex I nations have less reliable national statistics systems than most Annex I nations. These systems are the basis for emissions estimates, and experts noted that the estimates are only as good as the underlying data. For example, researchers estimated that the uncertainty of carbon dioxide emissions from China's energy sector was as high as 20 percent.[31] In contrast, reported uncertainties in estimates of carbon dioxide emissions from fossil fuel use in many developed nations are less than 5 percent. In addition, the International Energy Agency noted a relatively large gap between its energy statistics and those used in the national inventories of some non-Annex I nations, highlighting a need for better collection of data and reporting of energy statistics by some non-Annex I nations.[32]

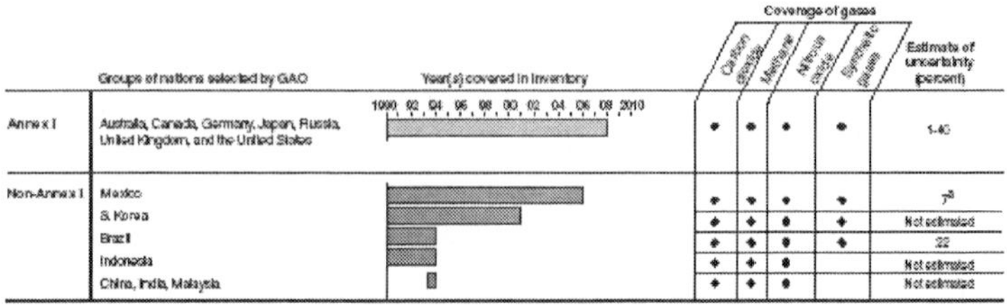

Source: GAO analysis of inventories.

Note: We selected the seven largest emitting Annex I nations and seven of the largest emitting non-Annex I nations

[a] This uncertainty estimate is from Mexico's 2006 submission. Mexico did not include an uncertainty analysis in its latest inventory submitted in 2009.

Figure 4. Comparison of Most Recent Greenhouse Gas Emissions Inventories Submitted by Selected Annex I and Non-Annex I Nations.

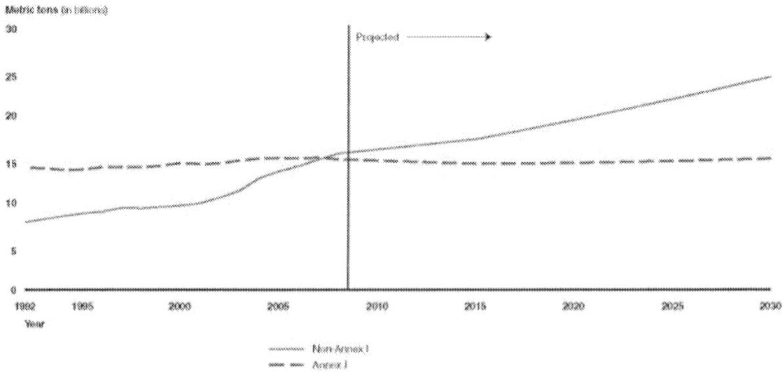

Source: GAO analysis of Energy Information Administration data.

Note: Does not include Monaco or Lichtenstein, which are both Annex I nations. These accounted for less than one percent of Annex I nations' emissions in 2007 according to inventory data submitted to the Convention Secretariat.

Figure 5. Energy-Related Carbon Dioxide Emissions in Annex I and Non-Annex I Nations, 1992 through 2030, Actual and Projected.

The lack of comparable, high quality inventories from non-Annex I nations is important because they represent the largest and the fastest growing portion of the world's emissions, and information about their emissions is important to international efforts to address climate change. (See figure 5.) For example, annual emissions from the seven selected non-Annex I nations grew by about 5 billion metric tons of carbon dioxide equivalent between 1990 and 2005, which was about the annual emissions of Canada, Germany, Japan, and

Russia in 2005 combined.[33] Recognizing the importance of information from non-Annex I nations, in March 2010, a National Research Council committee recommended that Framework Convention parties extend regular, rigorous inventory reporting and review to developing nations.[34]

Experts Identified Several Barriers to Improving the Comparability and Quality of Inventories

Experts we interviewed identified several barriers to improving the comparability and quality of inventories. First, 10 of the 12 experts who provided views about barriers said that a lack of data and scientific knowledge makes some types of emissions difficult to estimate for both Annex I and non-Annex I nations.[35] For example, current estimates of emissions related to biological processes, such as those from agriculture and land use, can be uncertain because of limited data. Specifically, nations do not always collect data on livestock nutrition, which can affect methane emissions. In addition, emissions related to some biological processes are difficult to estimate because they are not fully understood or are inherently variable. Emissions related to agriculture, for example, depend on the local climate, topography, soil, and vegetation. In March 2010, a National Research Council committee recommended further scientific research and data collection to reduce the uncertainties in estimates of agriculture, forestry, and land-use emissions.[36] Such emissions are important, contributing about one quarter of total global emissions in 2005, the most recent year for which global data were available. They are particularly important for some non-Annex I nations, where they can be the largest sources of emissions. In Brazil and Indonesia, for example, agriculture and land-use emissions accounted for about 80 percent of total emissions in 2005.

Second, 11 experts said that non-Annex I nations have limited incentives to produce better inventories. The current international system encourages Annex I nations with commitments under the Kyoto Protocol to improve their inventories.[37] This is because their ability to participate in the Kyoto Protocol's flexibility mechanisms—which provide a cost-effective way to reduce emissions—is linked to, among other things, the quality of certain aspects of their inventories.[38] Late submissions, omissions of estimates, or other shortcomings can all affect nations' eligibility to use these mechanisms.[39] Therefore, low-quality inventories can affect nations' ability to lower the costs of achieving their emissions targets. While four experts we interviewed said that this linkage between inventories and the flexibility mechanisms in the Kyoto Protocol has driven improvements in many Annex I nations' inventories, incentives for non-Annex I nations are limited. Furthermore, four experts said that some non-Annex I nations may avoid additional international reporting because they see it as a first step toward adopting commitments to limit emissions.

In addition, experts and the national communications of selected non-Annex I nations identified several other barriers to improving the quality and comparability of inventories from non-Annex I nations, including:

- *Less stringent reporting guidelines and lack of review.* Reporting guidelines differ between Annex I and non-Annex I nations. Non-Annex I nations do not need to annually submit inventories or to report on as many gases, for as many years, with as much detail, or in the same format as Annex I nations. They also do not have to follow all IPCC methodological guidelines, although they are encouraged to do so. Six experts said that this less stringent reporting regime has contributed to the lack of quality and comparability in inventories from non-Annex I nations. In addition, non-Annex I nations have not benefited from the feedback of technical reviews of their inventories, according to one expert.
- *Financial and other resource constraints.* Though eight experts generally said that many non-Annex I nations may lack needed financial and other resources, they differed on the magnitude and importance of additional international support. Non-Annex I nations may lack resources to improve data collection efforts, conduct additional research, or establish national inventory offices. The developed nations of Annex I provided the majority of about $80 million that has been approved for the latest set of national communications, which include inventories, from non-Annex I nations. However, one expert said that this has not been sufficient to fully support the activities needed. In their national communications, China and India indicated needing funding to, for example, improve data collection. Two experts said that improving non-Annex I nations' inventories may require significant resources. On the other hand, others said that the funds involved may be relatively small, or that financial constraints may not be significant, at least for major non-Annex I nations. For example, according to a report from a National Research Council committee, significant improvements in inventories

from 10 of the largest emitting developing nations could be achieved for about $11 million over 5 years.[40] While experts disagreed about the importance of additional funding, three said that international funding should support capacity development in each nation. They said that more continuous support would improve on the current, project-based method of funding, which encourages nations to assemble ad-hoc teams that collect data, write a report, and then disband.

- *Lack of data and nation-specific estimates of emissions factors.* According to four experts and the Convention Secretariat's summary of constraints identified by non-Annex I nations in their initial national communications, the lack or poor quality of data or a reliance on default emissions factors limit the quality of inventories. Most non-Annex I nations identified that missing or inadequate data was a major constraint for estimating emissions in at least one sector. For example, Indonesia reported that it did not estimate carbon emissions from soils because the data required were not available. Though inventory guidelines encourage the use of nation-specific emissions factors that reflect national circumstances, most non-Annex I nations use default values provided by the IPCC. The reliance on default values can increase uncertainties of estimates because national circumstances can differ significantly from the defaults. For example, Denmark's nation-specific emission factor for methane emissions from sheep is twice as large as the default. Thus, if Denmark had used the default value, it would have underestimated its emissions from sheep by half.

The Inventory Review Process for Annex I Nations Has Several Strengths and Some Limitations, and No Comparable Process Exists for Non-Annex I Nations

Experts said that the process for reviewing inventories from Annex I nations has several notable strengths. They also identified three limitations, which may present challenges in the future. Moreover, we found that although the review process includes steps to help ensure the quality of reviews, there is no independent assessment of the process' operations. Finally, there is no review process for inventories from non-Annex I nations.

The Process for Reviewing Inventories from Annex I Nations Has Several Strengths

Eight of the experts we interviewed said the process for reviewing inventories from Annex I nations has several notable strengths that enable it to generally meet its goals of providing accurate information on the quality of inventories and helping nations improve their inventories.[41] (Figure 6 below depicts the inventory review process.)

Experts identified four broad categories of strengths:

- *Rigorous review process.* Five experts said the rigorous review process gives them confidence that review teams can identify major problems with inventory estimates. For example, the Secretariat and review teams compare data, emission factors, and estimates from each inventory (1) from year to year, (2) with comparable figures in other inventories, and (3) with data from alternative sources, such as the International Energy Agency (IEA) and the United Nations Food and Agriculture Organization. Reviewers also ensure methods used to estimate emissions are appropriate and meet accepted guidelines. In addition, IEA officials inform the inventory review process by reviewing energy data in inventories and independently identifying issues for review teams to consider further.

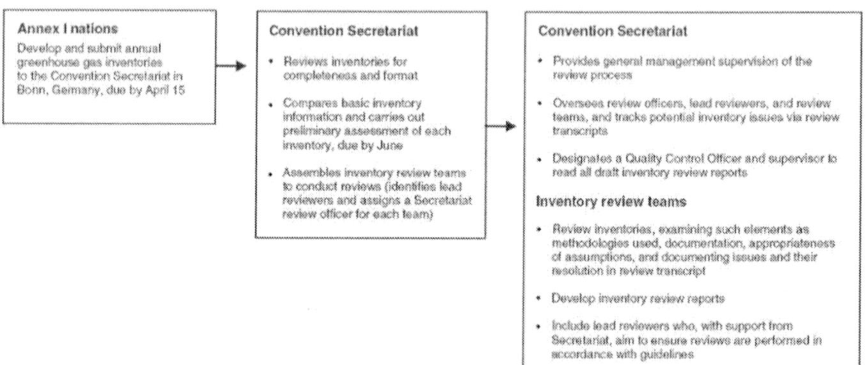

Source: GAO analysis of Convention documents and interviews with Secretariat officials.

Figure 6. Review Process for Inventories from Annex I Nations.

- *Qualified and respected reviewers.* Three experts we interviewed said that well-qualified and widely respected inventory reviewers give the process credibility. Secretariat officials told us that a relatively small number of people in the world have the expertise to evaluate inventories without further training. Parties nominate reviewers, including leading scientists and analysts, many of whom are also inventory developers in their home nations. Reviewers must take training courses and pass examinations that ensure they understand inventory guidelines and appropriate methodologies before serving on a review team. Two experts said reviewers' experience and qualifications allow them to assess the strengths and weaknesses in inventories, including whether nations use appropriate methodologies. This is particularly important because some nations use advanced or nation-specific approaches, which can be difficult to assess.
- *Capacity building.* Three experts said the inventory review process builds expertise among reviewers from developed and developing nations. Specifically, they said the review process brings inventory specialists together from around the world, where they learn from each other and observe how various nations tackle challenges in compiling their inventories. Two experts said that reviewers return home and can use the knowledge and contacts gained from their review team experiences to improve their national inventories.
- *Constructive feedback.* Two experts said that the inventory reviews provide constructive feedback to improve inventories from Annex I nations. This feedback includes identifying both major and minor shortcomings in inventories. Secretariat officials said that review teams, when they identify issues, must also offer recommendations for addressing them. For example, reviewers noted Russia's 2009 use of default assumptions for much of its uncertainty analysis, and recommended that Russia develop values that better match the methods and data used in making the emissions estimates.

For these and other reasons, three experts we interviewed said that the review process has helped improve the quality of inventories from Annex I nations. Secretariat officials said that when review teams point out discrepancies or errors, many nations revise and resubmit estimates to correct problems. For example, Australia revised its estimates of carbon dioxide emissions from croplands after a review team pointed out that changes in croplands management affect emissions. Australia's revisions decreased estimated emissions from croplands in 1990 by 138 percent, meaning the revisions had the effect of moving croplands from an estimated source of greenhouse gas emissions to a sink removing greenhouse gases from the atmosphere. For nations with Kyoto Protocol commitments, review teams may adjust estimates if they are not satisfied with a response to their findings.[42] For example, the team reviewing Greece's 2006 inventory concluded that estimates in several categories were based on methods, data, and emissions factors that did not adhere to reporting guidelines. The review team was not satisfied with Greece's response, and recommended six adjustments to Greece's estimates. These adjustments lowered Greece's official baseline energy sector emissions by 5 percent, from 82 million to 78 million metric tons of carbon dioxide equivalent.[43]

The Process for Reviewing Inventories from Annex I Nations Has Some Limitations

Experts, literature, and several nations identified some limitations of the review process, which may present challenges in the future if, for example, the process is expanded to incorporate non-Annex I nations. First, six experts we interviewed said the process does not independently verify emissions estimates or the quality of the underlying data. Review teams primarily ensure the consistency of inventories with accepted standards but do not check underlying activity data, such as the amount of fuel burned. Review teams do compare underlying data with those reported in other sources, but these other sources are not fully independent because they also come from the nations that supply the inventories. Two experts said that more thorough verification might involve comparing estimates to observed measurements or independently constructing estimates from raw data.

However, such approaches may be costly and, as a National Research Council committee reported, the other methods currently available do not allow independent verification of estimates. Furthermore, one expert said that the review of emissions estimates from agricultural soils and land-use sectors may be especially limited because of a lack of data and the inherent difficulty in measuring these emissions. The inability to more thoroughly assess inventories may reduce the reliability of review findings. For example, the inventory review process may have overlooked a significant shortcoming in at least one review. Specifically, in 2009, the national audit office of one Annex I nation found that its national inventory estimates may understate actual emissions by about a third because the inventory preparers used questionable statistics. The relevant agencies in that nation generally agreed with the audit office's recommendations based on its assessment. The review for that inventory, however, did not identify this issue.

Second, four experts we interviewed and several nations have expressed concerns about inconsistency across reviews, though the magnitude of this potential problem is unclear. The concerns relate to the potential for review teams to inconsistently apply standards when assessing an inventory. Secretariat officials said the process of reviewing inventories involves some degree of subjectivity, since reviewers use professional judgment in applying inventory review guidelines to a specific inventory. As a result, review teams might interpret and apply the guidelines differently across nations or over time. Four experts we spoke with, as well as several nations, have raised such concerns. For example, the European Community reported that some nations have received, on occasion, contradictory recommendations from inventory review teams.[44] Secretariat officials said lead reviewers are ultimately responsible for consistent reviews but that Secretariat staff assist the review teams during the process, and two Secretariat staff read through all draft inventory reports, in part to identify and resolve possible inconsistencies. In addition, lead reviewers develop guidance on consistency issues at annual meetings. The magnitude of this potential problem is unclear, in part because it has not been evaluated by an independent third party.

Third, three experts and officials we interviewed said there are not enough well-qualified reviewers to sustain the process. Three experts and Secretariat officials said that they did not know whether this shortage of available experts has affected the overall quality of reviews. The Secretariat has, in the past, reassigned staff and reviewers from work on national communications to the review of inventory reports, and it provides training to all reviewers to increase capacity and retain qualified reviewers. However, Secretariat officials said it may be difficult to sustain the quality of reviews in the future if the inventory review process is expanded to include inventories from non-Annex I nations without receiving additional resources, since this would substantially increase the demands on the review process.

The Review Process Does Not Have an Independent Assessment of Its Operations

The review process includes steps to help ensure the quality of reviews, but we found that its quality assurance framework does not independently assess the process. Secretariat officials said that lead reviewers oversee the drafting of review reports, and review officers, lead reviewers, and review teams maintain a review transcript to keep track of potential issues they have identified with inventories, of nations' responses to those issues, and of their resolution. However, lead reviewers, in the report of their 2009 meeting, expressed concern that these review transcripts are sometimes incomplete and are not always submitted to the Secretariat. In providing information on their experience with the review process and recommendations for improvements, the nations of the European Community suggested in late 2008 that the review process would benefit from establishing clear quality assurance and quality control procedures as well as from an annual analysis of its performance in relation to its objectives.[45] Secretariat officials said they designated a Quality Control Officer who, along with the supervisor of the review process, reads all draft review reports and may identify problems and check underlying information in reports. Furthermore, Secretariat officials said that lead reviewers meet annually to discuss the review process, assess and prepare guidance about specific issues or concerns about the review process, and develop summary papers to report to parties.

Nonetheless, the review process lacks an independent assessment of its operation. We examined several other review processes and found that periodic external assessments by independent entities can provide useful feedback to management and greater assurance that the review processes are working as intended.[46] Inventory guidelines call for Annex I nations to carry out quality assurance activities for their own inventories, including a planned system of reviews by personnel not directly involved in the process.[47] Though some United Nations and Framework Convention oversight bodies have the ability to assess the inventory review process, none have done so.[48] The Secretariat has internal auditors, but they have not audited the inventory review process and Secretariat officials said they did not know of any plans to do so. Although the Compliance Committee of the Kyoto Protocol has reviewed aspects of the review process, issuing a report with information on consistency issues, this report was not a systematic review and was not developed by people independent of the review process.

No Inventory Review Process Exists for Non-Annex I Nations

As stated earlier, inventories from non-Annex I nations do not undergo formal reviews. The Secretariat compiled a set of reports summarizing inventory information reported by non-Annex I nations, such as inventory estimates, national circumstances, and measures to address climate change.[49] However, Secretariat officials said they had not assessed the consistency of non-Annex I nations' inventories with accepted guidelines. These officials also said that they did not plan to compile another report covering non-Annex I nations' second inventories because the parties have not agreed to this. An expert we interviewed said that the quality of inventories from non-Annex I nations is unknown because their inventories have not been formally reviewed. Two experts said that some non-Annex I nations have resisted increased scrutiny of their inventories because of sovereignty concerns, meaning that nations do not want to disclose potentially sensitive information or data to other political bodies. The growth in greenhouse gas emissions along with lower quality inventories in some non-Annex I nations is likely to increase the pressure for a public review of their inventories in the future.

EXPERTS SAID THE INVENTORY SYSTEM IS GENERALLY SUFFICIENT FOR MONITORING COMPLIANCE WITH CURRENT AGREEMENTS, BUT FUTURE AGREEMENTS WITH NON-ANNEX I NATIONS COULD POSE CHALLENGES

Most experts we interviewed said that the inventory system for Annex I and non-Annex I nations is generally sufficient for monitoring compliance with current agreements. However, they said that the system may not be sufficient for monitoring non-Annex I nations' compliance with future agreements that include commitments for them to reduce emissions.

Experts Said the Inventory System Supports Current Agreements and International Negotiations

Eleven of the experts we interviewed said the inventory system—inventories and the process for reviewing them—is generally sufficient for monitoring compliance with current agreements, though five raised some concerns. All 11 of the experts who provided their views on the implications of the inventory system expressed confidence that inventories and the Convention's inventory review process are suitable for monitoring Annex I nations' compliance with existing commitments to limit emissions.[50] In part, this is because emissions in many Annex I nations primarily relate to energy and industrial activity, which can be more straightforward to estimate and monitor than emissions from land use and agriculture.

Nevertheless, five experts raised at least one of two potential challenges facing the current system. First, three said they were cautious until they see how the system performs under the more demanding conditions of submitting and reviewing inventories that will show whether nations have met their binding emission targets under the Kyoto Protocol.[51] When inventories are for years included in the Protocol's commitment period, nations may be more concerned about meeting emissions targets, and review teams may face pressure to avoid negative findings. Second, three experts said that flexibilities in the current inventory system or difficulties in measuring and verifying emissions from some agriculture and land-use segments could create complications for international emissions trading under the Kyoto Protocol. Emissions trading under the Kyoto Protocol allows nations with emissions lower than their Kyoto targets to sell excess allowances to nations with emissions exceeding their targets. Though parties to the Kyoto Protocol developed and agreed to the current system, three experts indicated that ensuring greater comparability of estimates between nations and types of emissions might be useful for emissions trading.[52] For non-Annex I nations, eight experts said that their lower quality inventories and lack of review do not present a current problem since these nations do not have international commitments to limit their emissions.

Seven of the experts said that the inventory system is sufficient to support international negotiations. To develop agreements, two experts said, negotiators need information on current and historic emissions from the nations involved. Annex I nations submit this information in their annual emissions inventories, the most recent of which cover emissions from 1990 to 2008. Although emissions estimates in most non-Annex I nations' inventories are outdated, seven experts said that there are enough independent estimates to provide negotiators with adequate information.[53] State officials said that independent estimates are useful, but official national inventories would be preferable because they can lead to more constructive discussions and can help create capacity in nations to better measure emissions. In international negotiations, State has emphasized the need for better information on emissions from all high-emitting nations, including non-Annex I nations.

Experts Said the Inventory System May Not be Sufficient for Monitoring Non-Annex I Nations' Compliance with Future Agreements, Though other Barriers Pose Greater Challenges

Different types of commitments would place different demands on the inventory system. Thus, the implications of the state of the inventory system for a future agreement will largely depend on the nature of that agreement. For Annex I nations, eight experts said that future commitments were likely to resemble current commitments and therefore the inventory system is likely to be sufficient. However, for non-Annex I nations, if future agreements include commitments to limit emissions, the current system is not sufficient for monitoring their compliance, according to nine experts. This is because non-Annex I nations do not submit inventories frequently, the quality of their inventories varies, and they do not undergo an independent technical review. Additional reporting and review could pose challenges since it could take time for non-Annex I nations to improve their inventories and Secretariat officials said that adding non-Annex I nations to the current inventory review process could strain the capacity of that system.

Some types of commitments by non-Annex I nations could be especially difficult to monitor and verify, according to experts. In the nonbinding 2009 Copenhagen Accord, many nations submitted the actions they intended to take to limit their greenhouse gas emissions, with Annex I nations committing to emissions targets for 2020 and non-Annex I nations announcing various actions to reduce emissions. Experts identified several challenges with monitoring the implementation of some of the actions proposed by non-Annex I nations (see table 3). For example, two experts said that monitoring emissions reductions from estimates of future business-as-usual emissions may prove challenging. They said this is because such actions may require parties to estimate reductions from a highly uncertain projection of emissions that would have otherwise occurred. parties would also have to develop and agree on guidelines to estimate and review business-as-usual emissions in addition to actual emissions. Similarly, monitoring reductions in the intensity of greenhouse gas emissions—emissions per unit of economic output, or gross domestic product—could pose challenges because of uncertainties in estimates of gross domestic product. One expert said that these challenges arise because the parties to the Convention created the current inventory system to monitor compliance and evaluate progress among Annex I nations with national targets. This expert added that parties to a new agreement may need to supplement the system to support the types of actions under consideration by non-Annex I nations.

Eight of the experts we interviewed said that parties to a future agreement could overcome or mitigate many of the challenges related to inventories. For example, two experts said that parties could design agreements that rely less on emissions estimates that are inherently uncertain or difficult to verify. For example, quantitative targets could apply only to sectors or gases that are relatively easy to measure and verify, such as carbon dioxide emissions from the burning of fossil fuels.

Three experts said that barriers other than the inventory system pose greater challenges to designing and reaching agreements on climate change. For example, nations disagree on the appropriate emissions limits for developed and developing nations. According to three experts, such disagreements were more of an obstacle to a comprehensive agreement in the latest round of negotiations in Copenhagen than were inventory issues. In addition, one expert pointed out that parties to international agreements generally have limited ability to get other parties to comply. For example, at least one nation with a binding emissions target under the Kyoto Protocol is unlikely to meet its target based on current inventory estimates and policies, according to this expert. Nations may be reluctant to agree to an international agreement until they have some assurance that other nations will follow through on their commitments.

CONCLUSIONS

High quality and comparable information on national greenhouse gas emissions is critical to designing and implementing international responses to climate change. The nations we reviewed meet their inventory reporting obligations, and review reports indicate this has resulted in generally high quality inventories from the seven highest emitting Annex I nations. However, the current inventory system does not request high quality emissions information from non-Annex I nations, which account for the largest and fastest growing share of global emissions. We found that the inventories from seven selected high emitting non-Annex I nations were generally outdated, not comparable, and of lower quality than inventories from Annex I nations.

Table 3. Select Actions Submitted under the Copenhagen Accord and Potential Challenges Identified by Experts

Type of action	Examples	Potential challenges identified by experts
Annex I nations		
National emissions targets	• 5-25 percent reduction from 2000 levels by 2020 (Australia)[a] • 17 percent reduction from 2005 levels by 2020 (Canada)[a] • 15-30 percent reduction from 1990 levels by 2020 (European Union, Japan, and Russia)[a] • In the range of a 17 percent reduction from 2005 levels by 2020 (United States)[a]	• Some issues to be resolved related to reporting land-use, land-use change, and forestry emissions. • Some concerns about using current inventories to support international emissions trading.
Non-Annex I nations—nationally appropriate mitigation actions		
Reductions from business-as-usual	• Up to 30 percent reduction from business-as-usual by 2020 (Mexico) • 30 percent reduction from business-as-usual by 2020 (South Korea) • 36-39 percent reduction from projected emissions by 2020 (Brazil)	• Would require new guidelines, reporting, and review for business-as-usual emissions projections. • Non-Annex I inventories are infrequent, less comprehensive, and are not reviewed. • Additional reviews could strain capacity of current system.
Sector-level targets and timetables	• Reduction in emissions related to deforestation (Brazil)	• May require new guidelines if sectors are not the same as defined in current guidelines. • Would require reporting and review of sector-level emissions.
Intensity targets	• 40-45 percent reduction in carbon dioxide emissions per unit of gross domestic product (GDP) by 2020 compared to 2005 (China) • 20-25 percent reduction in emissions intensity of GDP by 2020 compared to 2005 (India)	• Focus on forestry may present additional challenges. • Would require new guidelines, reporting, and reviewing for GDP figures. • Non-Annex I inventories are infrequent, less comprehensive, and are not reviewed. • Additional reviews could strain capacity of current system.
Policies and measures	• Promotion of energy efficiency (Indonesia)	• Would require new guide-lines, reporting, and review.

Source: GAO analysis based submissions to the Convention Secretariat and interviews with experts.

Note: This list represents challenges identified by one or more experts.

[a] All listed Annex I actions are conditional on actions by others, or, for the United States, on anticipated energy and climate legislation.

The existing gap in quality and comparability of inventories across developed and developing nations makes it more difficult to establish and monitor international agreements, since actions by both developed and developing nations will be necessary to address climate change under future international agreements. As a recent National Research Council committee study pointed out, extending regular reporting and review to more nations may require external funding and training, but the resources needed for the largest emitting developing nations to produce better inventories is relatively modest.

While our work suggests that the current inventory review process has notable strengths, we identified limitations that may present challenges in the future. For example, some experts and nations have reported concerns about inconsistent reviews and that resources may not be sufficient in the future. Stresses on the review process are likely to increase as review teams begin to review inventories that cover years in which some nations have binding emissions targets and if inventories from non-Annex I nations are subjected to inventory review under a future agreement. The Convention Secretariat has internal processes in place to help ensure quality reviews, but no systematic independent review to assess the merits of concerns about the consistency of reviews or to assess the need for additional qualified reviewers in the future. Addressing these issues could benefit the Secretariat by further enhancing confidence in its processes and ensuring that it has the resources necessary to maintain high quality reviews.

RECOMMENDATIONS FOR EXECUTIVE ACTION

We are making two recommendations to the Secretary of State:

1. Recognizing the importance of high quality and comparable data on emissions from Annex I and non-Annex I parties to the Convention in developing and monitoring international climate change agreements, we recommend that the Secretary of State continue to work with other parties to the Convention in international negotiations to encourage non-Annex I parties, especially high-emitting nations, to enhance their inventories, including by reporting in a more timely, comprehensive, and comparable manner, and possibly establishing a process for reviewing their inventories.
2. To provide greater assurance that the review process has an adequate supply of reviewers and provides consistent reviews, we recommend

that the Secretary of State, as the U.S. representative to the Framework Convention, work with other parties to the Convention to explore strengthening the quality assurance framework for the inventory review process. A stronger framework could include, for example, having an independent reviewer periodically assess the consistency of inventory reviews and whether the Secretariat has sufficient resources and inventory reviewers to maintain its ability to perform high quality inventory reviews.

AGENCY COMMENTS AND THIRD-PARTY VIEWS

We provided State, the Convention Secretariat, and EPA with a draft of this report for review and comment. State agreed with our findings and recommendations and said that the department has been working with international partners in negotiations and through bilateral and multilateral partnerships to support and promote improved inventory reporting and review.

The Convention Secretariat provided informal comments and said that it appreciated our findings and conclusions. The Secretariat said that the report provided a comprehensive overview of the existing system for reporting and reviewing inventories under the Convention and the Kyoto Protocol, as well as very useful recommendations on how this system could evolve in the future and steps to be taken to that end. The Secretariat noted our acknowledgement of the strengths of the inventory review process for Annex I nations. In addition, the Secretariat commented on our discussion of the limited availability of statistics against which to compare inventory data, saying that this lack of data does not imply that its review process lacks independent verification and that its review teams rely on available statistics in conducting their reviews. The Secretariat also said that the disparities in inventory quality across Annex I and non-Annex I nations should be viewed in the context of the "common but differentiated responsibilities" of developed and developing nations under the Convention.

In addition, EPA and the Convention Secretariat provided technical comments and clarifications, which we incorporated as appropriate.

John B. Stephenson
Director, Natural Resources and Environment

APPENDIX I. SCOPE AND METHODOLOGY

Our review provides information on: (1) the comparability, quality, and barriers to improving inventories submitted by developed and developing nations to the United Nations Framework Convention on Climate Change (the Convention); (2) the strengths and limitations of the Convention's inventory review process; and (3) the views of experts on the implications for agreements to reduce greenhouse gas emissions. To address all of these objectives, we reviewed relevant literature and Convention documents; met with officials from the Environmental Protection Agency (EPA), Department of State (State), the Convention Secretariat, and others to understand inventories, the inventory review process, and international negotiations; and summarized the views of experts on these issues.

Specifically, to address the first objective, we selected a nonprobability sample of 14 nations, seven Annex I nations—Australia, Canada, Germany, Japan, Russia, the United Kingdom, and the United States—and seven non-Annex I nations—Brazil, China, India, Indonesia, Malaysia, Mexico, and South Korea—based on the size of their emissions (including emissions from land-use and land-use change and forestry). We selected the largest emitting Annex I nations. For non-Annex I nations, we selected the largest emitting nations who had submitted inventories based on data available at the time.[54] We omitted Myanmar because it did not submit an inventory to the Convention. We also ensured coverage of major variations in selected nation's income and sectoral structure of their economies. To illustrate this variation, we used the World Bank's data on per capita income levels, and data from the World Resources Institute and Convention Secretariat on emissions from the energy and industrial processes sectors. The selected 14 nations represented about two thirds of the world's greenhouse gas emissions not related to land use and forestry in 2005. Our findings are not generalizable to other nations because the selected nations are not necessarily representative.

To assess the comparability and quality of inventories from Annex I nations, we summarized the results of the Convention's 2009 reviews of inventories from selected Annex I nations, the most recent reviews available. We did not independently assess the validity of data, assumptions, or methodologies underlying the inventories we reviewed. Though we identified some limitations with the inventory review process, we believe that reviews provide reasonable assessments of the comparability and quality of inventories from selected Annex I nations. For non-Annex I nations, we assessed whether the latest inventories from selected nations included estimates for all major greenhouse gases (carbon dioxide, methane, nitrous oxide, hydroflurocarbons, sulfur hexafluoride, and perfluorocarbons), for all sectors (energy, industrial processes, solvent and other product use, agriculture, land-use change and forestry, and waste) and various years, and checked for inclusion of key inventory characteristics, including descriptions of uncertainty and quality assurance and quality control measures, adequate documentation to support estimates, a comparable format, and analysis to identify emissions from key sources. Though inventory guidelines do not call for all of these from non-Annex I nations, we believe they are indicative of the quality and comparability of inventories. We did not independently assess emissions estimates from non-Annex I nations. We used the quality principles agreed to by parties for Annex I nations—transparency, consistency, comparability, completeness, and accuracy—as the basis of our review of all inventories and in our discussions with experts.[55] We also provide information on the reported uncertainty of emissions estimates, a more objective indicator of their precision, and on the timeliness of inventory submissions. To identify barriers to improving inventories, we reviewed relevant literature, including national communications from the seven selected non-Annex I nations, and summarized the views of our expert group.

To address the second objective, we summarized the results of semi-structured interviews with experts and Secretariat officials. We reviewed Convention documentation about the inventory review process, including Compliance Committee and Subsidiary Body for Implementation reports.

To address all three objectives, we summarized findings in the literature and the results of semi-structured interviews with experts. First, we identified 285 experts from our review of the literature and recommendations from U.S. and international government officials and researchers. From this list, we selected 15 experts based on (1) the relevance and extent of their publications, (2) recommendations from others in the inventory field, and (3) the extent to which experts served in the Consultative Group of Experts (a group assembled by the Convention to assist non-Annex I nations improve their national communications), as lead reviewers in the Convention's inventory review process, or were members of the National Research Council's committee on verifying greenhouse gas emissions. Finally, to ensure coverage and range of perspectives, we selected experts who had information about key sectors, like the agriculture and energy sectors, came from both Annex I and non-Annex I nations and key institutions, and provided perspectives from both those who were involved in the inventory review process and from those not directly involved in preparing or reviewing inventories. Appendix II lists the experts we interviewed, which included agency and international officials, researchers, and members of inventory review teams. We conducted a content analysis to assess experts' responses and grouped responses into overall themes. The views expressed by experts do not necessarily represent the views of GAO. Not all of the experts provided their views on all issues. We identify the number of experts providing views where relevant.

During the course of our review, we interviewed officials, researchers, and members of inventory review teams from State, EPA, and the Department of Energy in Washington, D.C.; the Convention Secretariat's office in Bonn, Germany; and from various think tanks, nongovernmental organizations, and international organizations.

We conducted this performance audit from September 2009 to July 2010 in accordance with generally accepted government auditing standards. Those standards require that we plan and perform the audit to obtain sufficient, appropriate evidence to provide a reasonable basis for our findings and conclusions based on our audit objectives. We believe the evidence obtained provides a reasonable basis for our findings and conclusions based on our audit objectives.

APPENDIX II. LIST OF EXPERTS

Scott Barrett, Columbia University

Clare Breidenich, Independent Consultant

Jane Ellis, Organisation for Economic Co-operation and Development

Thelma Krug, Inter American Institute for Global Change Research, Brazil

Bill Irving, Environmental Protection Agency

Bo Lim, United Nations Development Programme

Gregg Marland, Oak Ridge National Laboratory

Julia Martinez, Climate Change Program, Mexico

Sebastian Oberthür, Vrije University Brussels

Riitta Pipatti, Head of the Greenhouse Gas Inventory Unit at Statistics Finland

Helen Plume, Ministry for the Environment, New Zealand

Michael Prather, University of California, Irvine

Kristin Rypdal, Office of the Auditor General of Norway

Karen Treanton, International Energy Agency

David Victor, University of California-San Diego

End Notes

[1] The Convention's objective is to stabilize greenhouse gas concentrations in the atmosphere at a level that would prevent dangerous man-made interference with the climate system within a time frame sufficient to allow ecosystems to adapt naturally to climate change, to ensure that food production is not threatened, and to enable economic development to proceed in a sustainable manner. The five other gases are methane, nitrous oxide, and three synthetic gases—hydrofluorocarbons, perfluorocarbons, and sulfur hexafluoride.

[2] Under the Kyoto Protocol, each nation with a binding emissions target has a specified greenhouse gas emission limitation or reduction commitment.

[3] While the United States is a Party to the Convention and signed the Kyoto Protocol in 1998, it is not bound by the protocol's terms because it has not been ratified by the Senate.

[4] Human activities can cause the emission of greenhouse gases as well as their removal from the atmosphere. For example, activities that promote the growth of forests can lead to the removal of carbon dioxide from the atmosphere. Such removals are also called sinks. In this report, we use the term greenhouse gas emissions to refer to both emissions and removals unless otherwise noted.

[5] When the Convention and its annexes were agreed to in 1992, Annex I included all members of the Organization for Economic Cooperation and Development at that time, Russia, Baltic nations, and several Central and Eastern European nations. Annex I was subsequently amended and, in 1998, several other Central and Eastern European nations, as well as Luxembourg and Liechtenstein, were added. Although Kazakhstan is not listed in Annex I of the Convention, it has expressed its intent to be bound by the Convention's reporting requirements for Annex I nations and began submitting inventories in 2009. Additionally, in December 2009, the parties agreed to add Malta to the list of Annex I nations. This amendment will go into effect 6 months after the Convention Secretariat's communication to the parties about the amendment's adoption.

[6] The IPCC is a United Nations organization that, among other things, assesses the economic, scientific, and technical aspects of climate change.

[7] Our prior report looked at inventories from Germany, Japan, the United Kingdom, and the United States among Annex I nations, and from China, India, and Mexico among non-Annex I nations. GAO, *Climate Change: Selected Nations' Reports on Greenhouse Gas Emissions Varied in Their Adherence to Standards*, GAO-04-98 (Washington, D.C.: Dec. 23, 2003).

[8] See World Resources Institute, *Climate Analysis Indicators Tool (CAIT) version 7.0*, http://cait.wri.org (Washington, D.C., accessed May 25, 2010) and estimates from the Emission Database for Global Atmospheric Research in International Energy Agency (IEA), *CO_2 Emissions from Fuel Combustion, 2009 Edition* (Paris, France, 2009).

[9] Fossil-fuel related emissions are the largest contributor to global emissions, representing about two-thirds of global emissions in 2005. The Energy Information Administration is a statistical agency within the Department of Energy that collects, analyzes and disseminates independent information on energy issues. Energy Information Administration, *International Energy Statistics Database,* www.eia.doe.gov/emeu/international (Washington, D.C., accessed May 11, 2010) and *International Energy Outlook 2010,* DOE/ EIA-0484 (Washington, D.C., May 2010).

[10] EPA, *Inventory of U.S. Greenhouse Gas Emissions and Sinks: 1990-2008*, EPA-430-R-10-006, http://epa.gov/climatechange/emissions/usgginventory.html (Washington, D.C., Apr. 15, 2010).

[11] We report estimates of emissions not related to land use because of the lack of comparable data on land use emissions for many nations for many years. World Resources Institute, *CAIT version 7.0.*

[12] Because greenhouse gases differ in their potential to contribute to climate change, each gas is assigned a unique weight, called a global warming potential, based on its heat absorbing ability relative to carbon dioxide over a fixed period. This weighting provides a way to convert emissions of various greenhouse gases into a common measure, called carbon dioxide equivalent, which is used throughout this report.

[13] The Secretariat is also funded by certain fees, which generated nearly $77 million from 2008 to 2009.

[14] Various natural processes also emit greenhouse gases, such as forest fires and biological emission of methane from wetlands. Though differentiating emissions from human activities versus natural emissions can pose challenges, inventories under the Convention generally include emissions from human activities.

[15] Three IPCC guidelines are currently being used, each addressing different aspects of inventories: *Revised 1996 IPCC Guidelines for National Greenhouse Gas Inventories, Good Practice Guidance and Uncertainty Management in National Greenhouse Gas Inventories*, and *Good Practice Guidance for Land Use, Land-Use Change and Forestry.* The IPCC has recently updated its general inventory guidelines, but the parties have not yet agreed to start using these updated guidelines (*2006 IPCC Guidelines for National Greenhouse Gas Inventories*).

[16] Some nations are collecting more detailed information on emissions at major facilities. In the United States, the EPA recently issued a rule establishing mandatory greenhouse gas reporting requirements for owners and operators of certain facilities that directly emit greenhouse gases as well as for certain fossil fuel suppliers and industrial greenhouse gas suppliers. The first reports are due in 2011. In addition, about 10,500 electric power and industrial installations across 27 European nations have been calculating and reporting greenhouse gas emissions under the European Union Emission Trading System. Most of these nations indicated they have used this detailed emissions information in developing their 2010 inventories.

[17] The Convention, *Guidelines for the Preparation of National Communications by Parties Included In Annex I to the Convention, Part I: UNFCCC Reporting Guidelines on Annual Inventories*, FCCC/SBSTA/2006/9 (Bonn, Germany, Aug. 18, 2006).

[18] See http://unfccc.int/nationalreports/annexi_ghg_inventories/national_inventories_submissions/items/5270.php .

[19] Five Annex I nations with economies in transition—Bulgaria, Hungary, Poland, Romania, and Slovenia—are allowed to use other years as baselines.

[20] In its 2009 inventory, Germany used updated guidance and emission factors to estimate emissions from agriculture, but it did not justify the changes, as the parties have agreed to do. For 2006 emissions, the 2009 changes to methodology for agriculture reduced emissions by about 1.2 percent of the national total. The 2010 revisions have essentially reversed that.

[21] The 2010 U.S. inventory submission includes updated carbon content assumptions.

[22] EPA officials said they will include updated figures for the 2011 inventory submission.

[23] This is the statistical meaning of uncertainty. According to the IPCC, the term is sometimes used more generally and imprecisely when referring to greenhouse gas inventories—see IPCC, *IPCC Good Practice Guidance and Uncertainty Management in National Greenhouse Gas Inventories*, p. A3.19 (Geneva, 2000). An uncertainty percentage is often given as "plus or minus," meaning that the actual value could be either above or below the estimate by that percentage. Ranges reported here are for the 95 percent confidence interval.

[24] National Research Council, *Verifying Greenhouse Gas Emissions: Methods to Support International Climate Agreements* (Washington, D.C., National Academies Press, 2010).

[25] Of all 153 non-Annex I nations, only 24 have submitted more than their first inventories.

[26] Brazil and Indonesia estimated annual emissions for 1990 through 1994. These inventories were included in nations' national communications submitted on the following dates: Brazil, December 2004; China, December 2004; India, June 2004; Indonesia, October 1999; Malaysia, August 2000; Mexico, December 2009; and South Korea, December 2003.

[27] IEA, *CO2 Emissions from Fuel Combustion, 2009 Edition*.

[28] Mexico's quantitative uncertainty analysis was a part of its third National Communication, but not in the fourth National Communication.

[29] Unlike for Annex I nations, the reporting guidelines only encourage non-Annex I nations to use a common format.

[30] Mexico's key source analysis was a part of its third National Communication.

[31] Gregg, Jay S.; Andres, Robert J.; and Marland, Gregg, "China: Emissions Pattern of the World Leader in Carbon Dioxide Emissions from Fossil Fuel Consumption and Cement Production," *Geophysical Research Letters*, vol. 35 (2008).

[32] IEA, *CO_2 Emissions from Fuel Combustion, 2009 Edition*, I.5.

[33] Emissions for selected non-Annex I nations are from World Resources Institute, and from inventories submitted to the Secretariat for Canada, Germany, Japan, and Russia.

[34] National Research Council, *Verifying Greenhouse Gas Emissions*, 6.

[35] Some of the experts we interviewed did not provide views on barriers to improving inventories. Twelve experts provided the views discussed in this section.

[36] The committee recommended the production of global land-use and land cover maps every 2 years and the development of a research program to improve methods for estimating agriculture, forestry, and land-use emissions. See National Research Council, *Verifying Greenhouse Gas Emissions*, 9.

[37] Belarus, Turkey, and the United States do not have binding emissions targets or commitments under the Kyoto Protocol, though all are included in the Convention's Annex I.

[38] The Kyoto Protocol established three mechanisms that provide cost-effective ways for industrial nations to reduce their emissions. Emissions trading allows nations with emissions lower than their Kyoto targets to sell excess allowances to nations with emissions exceeding their targets. The Clean Development Mechanism and Joint Implementation allow nations with binding targets to implement projects that reduce or avoid emissions—such as the construction of renewable energy infrastructure—in developing nations that do not have binding emissions targets or industrialized nations respectively. These projects can earn credits, which industrial nation sponsors can use for compliance with their Kyoto targets.

[39] When inventory review teams find that a Kyoto Protocol Annex I nation's inventory is incomplete or not consistent with applicable guidelines, they may suggest "adjustments" to correct the shortcoming. Nations are ineligible if, for example, their inventory (1) is not submitted within 6 weeks of the due date, (2) omits an estimate for certain categories of emissions, or (3) annual adjusted emissions for a single year in the commitment period exceeds reported emissions by more than 7 percent. See, *Guidelines for the Preparation of the Information Required Under Article 7 of the Kyoto Protocol,* FCCC/KP/CMP/2005/8 Add.2, Decision 15/CMP.1, par 3.

[40] National Research Council, *Verifying Greenhouse Gas Emissions*, 6.

[41] Some of the experts did not provide views on the strengths and limitations of the inventory review process. Twelve experts provided the views discussed in this section.

[42] Because the authority for the review teams to make adjustments is provided by the Kyoto Protocol, the review teams cannot adjust estimates for Belarus, Turkey, or the United States, the Annex I nations that do not have Kyoto Protocol commitments.

[43] The Convention, *Report of the Review of the Initial Report of Greece*, FCCC/IRR/2007/GRC (Bonn, Germany, Dec. 28, 2007).

[44] France, "Submission by France on Behalf of The European Community and Its Member States," *Experiences With and Lessons Learned From the Review of Initial Reports Under the Kyoto Protocol, Including Recommendations for Improvements*, FCCC/SBI/2008/MISC.7 (Bonn, Germany, 2008) 7.

[45] France, *Experiences With and Lessons Learned From the Review of Initial Reports,* 7.

[46] Though neither the Secretariat nor review teams are required to follow them, we examined the standards, descriptions or guidelines of the review processes in the following: (1)

International Standard Organization (ISO) guidance for verifying greenhouse gas assertions (ISO, *Greenhouse gases—Part 3: Specification with Guidance for the Validation and Verification of Greenhouse Gas Assertions*, ISO 14064-3 (Geneva, Switzerland, 2006).), (2) the study process used by the National Academies, (3) Government Auditing Standards (GAO, *Government Auditing Standards July 2007 Revision*, GAO-07-731G (Washington, D.C.: July, 2007).), and (4) the Institute for Internal Auditors standards (The Institute of internal Auditors, *International Professional Practices Framework* (Altamonte Springs, Florida, January 2009).).

[47] IPCC, "Quality Assurance and Quality Control," *Good Practice Guidance and Uncertainty Management in National Greenhouse Gas Inventories* (2000).

[48] The Framework Convention's Subsidiary Body for Implementation assists the parties in assessing and reviewing the effective implementation of the Convention, and the Subsidiary Body for Scientific and Technological Advice provides the parties with information and advice on scientific and technological matters, including matters that relate to inventories.

[49] The Convention, *Sixth Compilation and Synthesis of Initial National Communications from Parties Not Included in Annex I to the Convention*, FCCC/SBI/2005/18 (Bonn, Germany, 2005).

[50] Some of the experts we interviewed did not provide views on the implications for international agreements because, for example, they felt they were not experts in these issues. Eleven experts provided views discussed in this section.

[51] Inventories that Annex I parties with Kyoto commitments submitted in 2010, which are currently being reviewed, are the first to include emissions estimates for years included in the Kyoto Protocol's 2008-2012 commitment period, though inventories to establish baseline emissions have already been reviewed.

[52] Under the Kyoto Protocol, nations can only use certain emissions from land-use change and forestry to meet the Protocol's targets.

[53] For example, IEA publishes yearly estimates of carbon dioxide emissions from the combustion of fossil fuels, the most recent of which includes estimates from 1971 to 2007 for 140 nations. In addition, the Emission Database for Global Atmospheric Research, developed by the European Commission's Joint Research Centre and the Netherlands Environmental Assessment Agency, estimates other sources of carbon dioxide and other greenhouse gases emissions. See IEA, *CO2 Emissions from Fuel Combustion, 2009 Edition* and http://edgar.jrc.ec.europa.eu/ .

[54] Based on more recent data, Malaysia is no longer one of the seven highest emitting non-Annex I nations.

[55] The Convention, *Guidelines for the Preparation of National Communications by Parties Included in Annex I to the Convention, Part I: UNFCCC Reporting Guidelines on Annual Inventories,* FCCC/SBSTA/2006/9 (Bonn, Germany, Aug. 18, 2006).

In: Greenhouse Gas Emissions
Editors: M. Knox and J. D. Mark

ISBN 978-1-62081-287-7

Chapter 3

GREENHOUSE GAS EMISSIONS: PERSPECTIVES ON THE TOP 20 EMITTERS AND DEVELOPED VERSUS DEVELOPING NATIONS*†

Larry Parker and John Blodgett

SUMMARY

Using the World Resources Institute (WRI) database on greenhouse gas emissions and related data, this report examines two issues. The first issue is the separate treatment of developed and developing nations under the United Nations Framework Convention on Climate Change (UNFCCC), the Kyoto Protocol, and the Copenhagen Accord. This distinction has been a pivotal issue affecting U.S. climate change policy. The second issue is the difficulty of addressing climate change through limiting greenhouse gas emissions to a specified percentage of baseline emissions (typically 1990). The data permit

* A version of this chapter also appears in *International Climate Change Negotiation and Investment*, edited by Robert A. Tamm, published by Nova Science Publishers, Inc. It was submitted for appropriate modifications in an effort to encourage wider dissemination of research.

† This is an edited, reformatted and augmented version of a Congressional Research Service publication, RL32721, January 7, 2010.

examination of alternative approaches, such as focusing on per capita emissions or the greenhouse gas emission intensity (measured as emissions per unit of economic activity). Key findings include:

- A few countries account for most greenhouse gas emissions: in 2005, China led by emitting 19% of the world total, followed closely by the United States with 18%; no other country reached 6%; the top eight emitters (those emitting 2% or more of total emissions) accounted for 58% of the 185 nations' emissions.
- Land-use effects (e.g., deforestation) on emissions are negligible for most nations, but they cause emissions to rise sharply for certain developing nations, most notably Brazil and Indonesia.
- While countries whose economies are dominated by oil- and gas-production have the highest per capita greenhouse gas emissions, in general developed nations rank high in per capita emissions (in 2005, Australia, the United States, and Canada ranked 6, 9, and 10, respectively, in the world), while developing nations tend to rank low (China, Brazil, Indonesia, and India ranked 81, 84, 117, and 148, respectively).
- The greenhouse intensity of the economy—the metric by which the George W. Bush administration addressed climate change, and by which China has proposed to set its objectives under the Copenhagen Accord—varies substantially among developed countries (in 2005, not accounting for land use, Ukraine emitted 512 tons/million international $GDP, while France emitted 80 tons/million $GDP, with the United States at 153 tons/million $GDP; developing nations range from the 136 (Mexico) to 372 (China).
- The time frame adopted for defining the climate change issue and for taking actions to address greenhouse gas emissions has differential impacts on individual nations, as a result of individual resource endowments (e.g., coal versus natural gas and hydropower) and stage of economic development (e.g., conversion of forest land to agriculture occurring before or after the baseline).

Differentiating responsibilities between developed and developing nations—as the UNFCCC does—has failed to engage some of the largest emitters effectively. Moreover, many developed countries have not achieved stabilization of their emissions despite the UNFCCC. Given the wide range of situations illustrated by the data, a flexible strategy that allows each country to play to its strengths may be necessary if diverse countries like the United States and China are ever to reach agreement. The difficulty in finding a common strategy was evidenced by the outcome of the Copenhagen meeting, which set a climate change objective of holding global warming to less than 2 degrees C but then left up to each country the choice of how to address emissions.

INTRODUCTION

Climate change is a global issue;[1] however, greenhouse gas emissions data on a global basis are incomplete. Some developing countries have no institutions for monitoring greenhouse gas emissions and have never reported such emissions to the United Nations Framework Convention on Climate Change (UNFCCC).[2] In a similar vein, data on individual greenhouse gases, sources, and land-use patterns vary greatly in quality. Despite shortcomings in the data, the emerging picture of emissions has implications for considering alternative policies for controlling emissions. First, the picture outlines the estimated contributions of individual countries. Second, evaluating those emissions in terms of socio-economic characteristics (e.g., population and economic activity) provides insights on the potentially divergent interests of differing groups of nations—especially concerning developed nations versus developing ones.[3]

The World Resources Institute (WRI) has compiled greenhouse gas emissions and related data from a variety of sources into a database that is available for analysis.[4] Covering 185 nations (plus a separate entry combining the members of the European Union),[5] the database includes total emissions, per capita emissions, and greenhouse gas (or carbon) intensity;[6] selected socio-economic indicators; and other measures. Emissions data for all six greenhouse gases[7] identified by the UNFCCC are available for 1990, 1995, 2000, and 2005 for both developed and non-Annex I nations. Data for carbon dioxide (CO2) are available back to 1850 and up to 2006 for both developed and non-Annex I nations. Data on the effects of land use change and forestry on CO2 emissions are only available from 1990 to 2005, and only for a subset of nations.

This report uses the data compiled by WRI to examine a pivotal and long-running issue surrounding U.S. climate change policy: the appropriate roles of developed and developing countries in addressing climate change.

The UNFCCC states as its first principle in Article 3:

> The parties should protect the climate system for the benefit of present and future generations of humankind, on the basis of equity and in accordance with their common but differentiated respon-sibilities and respective capabilities. Accordingly, the developed country parties should take the lead in combating climate change and the adverse effects thereof.[8]

U.S. policymakers have struggled with the "common but differentiated responsibilities" of all nations and with the pledge for the developed countries to "take the lead in combating climate change.... " Under the UNFCCC and the subsequent Kyoto Protocol, common actions include the responsibility to monitor and report emissions; differentiated actions include the commitment to reduce emissions for designated developed nations (including the United States), listed on Annex I to the UNFCCC (and hence known as Annex I nations).

The original UNFCC commitment was voluntary, and many Annex I nations, notably including the United States, failed to meet the objective of reducing 2000 emissions to a 1990 baseline.[9] Thus the Kyoto Protocol made mandatory individual Annex I nations' commitments of percentage reductions for 2008-2012, but meeting them has proved difficult – and the United States refused to join the commitment.[10] Under both the UNFCCC and the Kyoto Protocol, non- Annex I nations would be exempt from these specified control requirements—although they could voluntarily join in. This split in responsi-bilities—with the consequent lack of greenhouse gas control requirements for major emitting non-Annex I countries—played a key role in the United States' refusal to agree to the Kyoto Protocol.

Two key issues emerged from the UNFCC and Kyoto commitments to reduce emissions by developed nations: first, meeting the commitments is proving to be both technically and politically difficult; and second, it has become increasing evident that any reductions achieved by Annex I nations could be nullified by increases in emissions from non-Annex I nations like China and India that have been undergoing rapid economic growth and emitting increasingly large amounts of greenhouse gases – such that by 2005 China passed the United States to become the number one emitter of greenhouse gases in the world.

Justifications for the differential treatment of the developed, Annex I nations compared to the developing nations are both environmentally and economically based.

- Environmentally, the developed, Annex I nations have dominated emissions. Cumulatively, from 1850 to 2006, Annex I nations had emitted approximately 74% of energy-related CO2, while non-Annex I nations had contributed 24%.[11] In 1990, when the UNFCCC was being conceived, Annex I nations accounted for 60% of emissions of all six greenhouse gases, while the non-Annex I nations accounted for 40%. By 2005, however, non-Annex I nations dominated, accounting for 51% of total emissions, while Annex I nations accounting for approximately 47%. Thus, while Annex I nations still dominate cumulative emissions, the fact that non-Annex I nations are now contributing more than half the emissions confounds the assignment of future obligations.
- Economically, as the UNFCCC explicitly recognizes, the economic development being pursued by the non-Annex I nations depends importantly on expanded use of energy, including fossil fuels, which are the main source of carbon dioxide, the dominant greenhouse gas. From this perspective, a logic for the differing treatment of the two groups is that the developed, Annex I countries can afford to control emissions because they have achieved a relatively high standard of living, while the developing nations have the right and should have the opportunity to expand energy use as necessary for their economic development.

This distinguishing of the responsibilities of the Annex I and non-Annex I nations generates crucial and interrelated tensions:

- First, this approach means that Annex I nations bear the preponderance of the direct economic costs for addressing global climate change;
- Second, non-Annex I nations retain the opportunity to develop their economies using least-cost energy regardless of greenhouse gas emissions; this in turn means that from the perspective of the Annex I nations, at least some developing nations—which may be competing in certain economic sectors—appear to be getting a free ride;
- And third, despite investments in controls and resulting tensions between competing economies, actual global emissions will continue to rise if the increase in emissions from non-Annex I nations exceeds any decrease in emissions achieved by Annex I ones.

The crux of the Copenhagen Conference, to plot a post-Kyoto course for addressing climate change, was how to engage the two largest emitters, the United States and China – the former having rejected Kyoto in part because developing nations were not obligated to curtail emissions; and the latter having become the world's largest emitter of greenhouse gases. Politically, while George W. Bush administration had been a reluctant partner in the UNFCC process, including early negotiations pointing toward Copenhagen, President Obama has been a vigorous proponent of engagement. At the Copenhagen Conference, he met twice with Chinese Premier Wen Jiabao in an effort to move the negotiations forward.

The Copenhagen outcome showed both some progress in bridging the gap between the developed and developing nations, and continuing difficulties in finding common ground on how to reduce greenhouse gas emissions. The accord[12] did not mandate specific reductions, but set a goal of reducing global emissions "so as to hold the increase in global temperature below 2 degrees C, and take action to meet this objective consistent with science and on the basis of equity." Annex I nations commit to implement "quantified economy-wide emissions targets for 2020" and non- Annex I nations commit to implement "mitigation actions." Both sets of nations commit to reporting and verification procedures "in accordance with guidelines adopted by the Conference of the Parties." (Monitoring, reporting, and verification were a key demand of the United States of developing nations.) Also, the accord contained the promise of $100 billion a year by 2020 "to address the needs of developing countries."

To clarify how nations' emissions levels intersect with social and economic contexts, this paper focuses on the 20 individual nations that emitted the most greenhouse gases in 2005 (see **Appendix A**, **Appendix B**, and **Appendix C**). [13] In 2005, not taking into account emissions implications of land use and forestry, the top 20 represented about 75% of global greenhouse gas emissions—up slightly from about 73% in 1990 (latest available data from CAIT for all six greenhouse gases). In addition, data for the 27-member[14] European Union are included, as the Kyoto Protocol allows the EU to address its greenhouse gas emission obligations collectively. In 2005, the 27-nation EU was the third-largest emitter of greenhouse gases, after China and the United States.

A LOOK AT THE HISTORIC DATA

2005 and 1990 Emissions Data (without accounting for land use and forestry emission effects)

A compelling fact to emerge from the database is that a few countries account for most of the emissions. **Appendix A**, **Appendix B**, and **Appendix C** presents data concerning the top 20 greenhouse gas-emitting nations in 2005. They accounted for 75.3% of global emissions. Excluding land use data, by CAIT's accounting, China led in emitting greenhouse gases (1,974 million metric tons of carbon equivalent, MMTCE)[15] at 19.1% of the total, followed by the United States (1,892 MMTCE) at 1 8.3%.[16] No other country reached 6% of total emissions (although the collective 27-member EU accounted for 13.4%); overall, only eight countries emitted 2% or more. These top eight emitters accounted for 5 8.3% of global emissions and the next 13 top emitters accounted for another 17% of emissions.

Thus one implication of these data is that greenhouse gas control in the short term depends mainly on the actions of a relatively few nations; if the top 20 emitters[17] (or even the top eight) all acted effectively, the actions of the remaining 160-plus nations would be of relatively little import, at least for years.

A second compelling fact about those top emitters is that they are highly diverse and represent very different situations.[18] The top 20 nations include:

- Developed (Annex I) nations whose emissions *grew* between 1990 and 2005: the United States, Japan, Canada, Italy, Australia, France, Spain, and Turkey (ranked 2, 5, 8, 13, 14, 16, 18, and 20, respectively). These eight nations accounted for 30.5% of global greenhouse gas emissions in 2005.
- Developed (Annex I) nations whose emissions *declined* between 1990 and 2005, largely as a result of the collapse of the Eastern European and USSR socialist economies during the 1990s: Russian Federation, Germany,[19] and Ukraine, (ranked 3, 7, and 17, respectively). These three nations accounted for 9.0% of global greenhouse gas emissions in 2005.[20]
- Developed (Annex I) nations with free-market economies whose emissions *declined* between 1990 and 2005, largely because of a combination of low population growth, modest economic growth, and

the displacement of high- emitting fuels (coal) with alternatives: the United Kingdom (ranked 9), is the only member of this category.[21] It accounted for 1.7% of global greenhouse gas emissions in 2005.

- Developing (non-Annex I) nations, all of whose emissions rose during the period: China, India, Brazil, Mexico, Indonesia, South Korea, Iran, and South Africa (ranked 1, 4, 6, 10, 11, 12, 15, and 19, respectively). These eight nations accounted for 34.1% of global greenhouse gas emissions in 2000.

For the year 2005, then, 12 of the top 20 countries were Annex I countries, including 6 of the top 10 emitters. In 2005, the top 20 Annex I countries accounted for about 55% of the top 20 group's greenhouse emissions, compared with 45% for the developing, non-Annex I countries; in 1990, the relative shares were 69% and 31%, respectively, so the developing countries have been proportionately increasing their share.

Highlighting the tension between Annex I and non-Annex I perspectives, the number-one emitters of each group were the top two emitters overall: At the top were the leading developing, non-Annex I country, China; and the leading developed, free-market economy, the United States. Combined, these two countries alone accounted for 37.4% of total global emissions.

Longer-Term Historical Data (1850-2005)

The impact of emissions on climate change is believed to be cumulative over decades and even centuries. Thus a longer-term examination of data provides an important perspective, and is one reason for the differing treatments of the Annex I and non-Annex I nations. Available data give emissions estimates of energy-related CO_2 emissions back from1850 to 2006 (see **Appendix A** and **Appendix C**).

This longer-term view of emissions underscores the contribution of the Annex I nations:

- For all nations, excluding land use changes and forestry practices, Annex I countries' share of energy-related CO_2 emissions[22] over the period 1850-2005 is 74%; non-Annex I countries' share is 24% (see also **Table 1**).
- The relative rankings of several developing countries, including Brazil, South Korea, Indonesia, and Iran, drop substantially using a

longer historical baseline for emissions: from the 2005 rank to the 1850-2005 cumulative rank for CO_2, from 6th to 21st, 15th to 20th, 11th to 25th, and 12th to 23rd , respectively.

Greenhouse gas emissions, particularly energy-related emissions, are closely tied to industrialization. As "developed" is considered by many to be synonymous with "industrialized," it is not surprising that the developed countries dominate cumulative emissions, while developing ones are increasing their current annual share.

Impact of Land Use

Changes in land use can significantly affect net levels of emissions.[23] In general, deforestation increases CO_2 emissions and afforestation decreases them. Certain agricultural practices can increase emissions of methane or nitrous oxide; other agricultural processes can sequester carbon. However, data on the effects on emissions of land use changes and forestry practices, and their conversion into equivalent units of greenhouse gas emissions, are both less available and less robust than data on emissions. Therefore, this discussion is at best illustrative (see **Appendix A** and **Appendix C**; note that numerous countries lack data on land use and forestry).

Including land use in the calculations for 2005 focuses discussion on certain developing countries.

- Land use changes and forestry practices in certain developing countries, notably Brazil and Indonesia, are having the effect of substantially upping their relative emissions ranks. Counting land use, Brazil's emissions in 2005 rise from 276 MMTCE to 776 MMTCE (+181%), and Indonesia's rise from 159 to 557 (+250%). This ups their rankings of total emissions in 2005 from 6th to 3rd , and 11th to 4th, respectively.
- Compared to Brazil and Indonesia, the impact of accounting for land use on other top 20 emitters is much less. The next biggest adjustment is for Mexico, whose emissions rise 6% when land use is accounted for. For the United States, net emissions drop by 32 MMTCE (-1.7%); its relative rank (as number 2 in 2005) does not change when land use is taken into account.

Historic land use and forestry data are not available. Evaluating the impact of land use and forestry at any one time directs attention to those few countries undergoing particular points in the development cycle. For many countries, land-clearing and agricultural development occurred long ago: the Western developed nations and China and India, for example, have long-established agricultural practices; in contrast, Brazil and Indonesia have over the past few decades been clearing large regions of forest and jungle for timber and/or conversion to agriculture, releasing greenhouse gases (or removing sinks). In terms of the UNFCCC and the Kyoto Protocol, and potentially the Copenhagen Accord, including land use in the equation for controlling emissions disadvantages certain countries whose exploitation of resources and development of agriculture are occurring at a particular moment in history, while for other countries the effects of past changes in land use and forestry practices are embedded in their baseline emissions.

Implications of Focusing on Emissions Levels for International Actions

The data on greenhouse gas emissions highlight issues of both effectiveness and fairness in the effort to address global climate change. Differentiating responsibilities between Annex I and non- Annex I nations, as the UNFCCC has, does not focus efforts on all of the largest emitters. As **Table 1** shows, the emissions dominance of Annex I nations that existed in 1990 has ended: in 2005 non-Annex I nations' global greenhouse gas emissions definitively surpass those of Annex I nations, by a margin of 15% when taking land use and forestry into account. On the other hand, on the basis of energy-related CO_2 emissions, cumulative from 1850-2006, Annex I nations still dominate by margin of 3 to 1.

Moreover, contradictory issues of fairness arise. For Annex I countries, the present scheme of controlling greenhouse gases requires them to bear essentially all the direct economic costs. For non-Annex I countries, to the extent that development is linked to increasing greenhouse gas emissions, imposing controls on them could slow their development and hold down their standards of living vis-a-vis the developed nations.

Table 1. Shares of Global Emissions by the Industrialized (Annex I), Developing (non-Annex I), and Top 20 Countries.

Indicator	Industrialized (Annex I) Countries n = 38[a]	Developing (non-Annex I) Countries n = 147	Top 20 Nations in 2005
1990 GHG Emissions (excl. land use)	59.9%	39.8%	75.2%
2005 GHG Emissions (excl. land use)	47.0%	51.5%	75.3%
1990 GHG Emissions (with land use)	50.1%	47.8%	72.6%
2005 GHG Emissions (with land use)	41.2%	56.6%	73.6%
Cumulative Energy-Related CO_2 Emissions 1850-2006 (excl. land use)	74.4%	24.5%	83.0%

Source: CRS calculations; Climate Analysis Indicators Tool (CAIT) Version 7.0 (Washington, DC: World Resources Institute, 2009).

a. Counting the European Union countries individually, excluding the EU as a collective member.

Finally, the focus on emissions levels at specific times (e.g., a baseline of 1990) has differential and arbitrary impacts on individual nations.

- Looking at the industrialization process, to the extent that fossil fuel use is a necessary ingredient of economic development, as acknowledged by the UNFCCC, the emergence of the global climate change issue at this time effectively determines the distinction between the developed, Annex I nations and the developing, non-Annex I nations. For Annex I nations, that energy exploitation has been incorporated into their economies and is part of their baseline for considering any controls on greenhouse gases. For developing, non- Annex I nations, however, economic development will require expanded energy use, of which fossil fuels can be the least costly. Thus imposing limits on fossil energy use at this time could result in developing countries being relegated to a lower standard of living than those nations that developed earlier.

- Similarly, certain land-use activities, such as clearing land for agriculture and exploiting timber, affect net greenhouse gas emissions. Nations that are currently exploiting their resource endowments, such as Brazil and Indonesia, could find themselves singled out as targets for controls. Yet developed nations, like the United States and most European countries, which exploited such resources in the past, have those greenhouse gas implications embedded in their baselines.

- Also, the focus on 1990 as a baseline means that the Eastern European and former Soviet Union nations have the advantage of reductions in emissions from their subsequent economic contractions, which will allow them room for growth. Likewise, the discovery and exploitation of North Sea gas has allowed Great Britain to back out coal and thereby reduce emissions since the baseline.

In all these cases, the time frame adopted for defining the climate change issue and for taking actions to address greenhouse gas emissions has differential impacts on individual nations, as a result of their individual resource endowments[24] and stage of economic development. The differential impacts give rise to perceived inequities. Thus the effort to find a metric for addressing greenhouse gas emissions baselines and targets that will be perceived as equitable is challenging.

Alternative Perspectives

The problems raised above prompt the question: What alternatives to controls derived from historically based emissions levels are available?

Alternative metrics for taking into account greenhouse gas emissions and economic development include per capita emissions and economic intensity of emissions.[25]

Per Capita Emissions

The socioeconomic differences between the developed, Annex I nations and the developing nations lead to considerations about emissions other than simply their absolute amounts. One alternative is to consider per capita emissions: All else equal, populous nations would emit more greenhouse gases than less populated ones. On this basis, the difference between developed, Annex I countries and non-Annex I ones is apparent.

Appendix A and **Appendix B** show that of the top 20 emitters in 2005, the highest ranked by per capita greenhouse gas emissions[26] are developed countries (Australia, United States, and Canada, ranked 6, 9, and 10, respectively). Their per capita emissions (7.5, 6.4, and 6.2 tons per person, respectively) are nearly double the emissions of the highest-ranked developing country in the top 20 (South Korea, at 3.2), and over four times that of China (1.5). The rankings for the non-Annex I countries in the top 20 emitters range from 29 (South Korea) to 148 (India), with China ranked 81. In contrast, Annex I countries range from 6 (Australia) to 51 (France), with the United States at 9. Reasons the United States, Australia, and Canada are so high on this measure include their dependence on energy-intensive transport to move people and goods around countries of large size and relatively low population density, the use of coal for power generation, and the energy requirements for resource extraction industries.

Thus, if one were considering how to control greenhouse gas emissions, one way of trying to bridge the different interests of the developed, Annex I nations and the developing ones would be to focus on per capita emissions as a way of giving each nation an equitable share of energy use. For the United States compared to the developing world, this metric could imply constraints, depending on the compliance time frame and future technological advancements. Likewise, this approach could permit most less-developed countries to increase their emissions to accommodate expanding economies.

Greenhouse Gas Intensity of Economy

Another alternative for evaluating a nation's contribution to greenhouse gas emissions is to consider how efficiently that nation uses energy (and conducts other greenhouse gas-emitting activities) in producing goods and services. This concept is captured by greenhouse gas intensity—or carbon intensity[27]—measured as the amount of greenhouse gases emitted per million dollars of gross domestic product, measured in international dollars (parity purchasing power) (see **Appendix A** and **Appendix C**). Carbon intensity as a greenhouse gas indicator has received considerable attention since President Bush decided to use it as a benchmark for his voluntary climate change program. Also, the World Resources Institute has advocated its use as an appropriate index for developing, non-Annex I nations.[28]

A nation's greenhouse gas intensity reflects both its resource endowment and the energy- intensiveness of its economy. In terms of energy resources, countries with rich resources in coal would tend to be higher emitters, while countries with rich resources in hydropower or natural gas would tend to be lower emitters. In terms of economic activity, countries with major heavy industry, major extractive industries, and extensive transportation systems tend to be higher emitters, while countries without these and/or dominated by service industries would tend to be lower emitters. As noted in terms of emissions, taking into account land use sharply increases the greenhouse gas intensity of Brazil and Indonesia.

The top 20 emitters in 2005 (see **Appendix A** and **Appendix C**) range widely in greenhouse gas intensity: from 512 tons per million international $GDP (Ukraine, which relies heavily on coal) to 80 tons/million international $GDP (France, which relies heavily on nuclear power for generating electricity). (The larger the intensity number, the more GHGs emitted per dollar of GDP: from a climate change perspective, the lower the intensity the better.) These are both Annex I nations; non-Annex I nations have a narrower range, from the 136 tons/million international $GDP (Mexico) to 372 tons/million international $GDP (China). Taking into account land use, however, would dramatically raise the intensity of Brazil and Indonesia: in 2005 it jumped Brazil by 182%, to 490 tons/million international $GDP and Indonesia by 250%, to 790 tons/million international $GDP; the next largest increase from land use change was Mexico at 6%.

As a metric for considering how to control greenhouse gas emissions, intensity has an inherent political appeal: for most nations, intensity is declining. For the world, greenhouse gas intensity declined at a rate of -1.6%

annually from 1990 to 2005. Causes of this decline in intensity are cost efficiencies that focus attention on the efficient use of energy, policies promoting the use of alternatives to fossil fuels generally and to coal in particular, and substitutions for perfluoro- and hydrofluorocarbons. A consequence of focusing on intensity, however, is that even with declining intensity, actual emissions can rise – as a result of population growth and economic growth.[29]

For greenhouse gas intensity, in 2005 the United States ranked number 122 in the world, making this a more favorable metric than absolute emissions (the United States ranked number 2 in the world) and per capita emissions (the United States ranked number 9). Of the indicators examined here, the United States gets the most favorable results from this one. Nevertheless, in absolute terms, the United States is relatively less efficient with respect to intensity compared with Western European countries (the EU-27 would have ranked 154) and Japan ranked 166. In addition, the United States is less efficient than non-Annex I emitters South Korea and Mexico, but it is more efficient than China, India, Brazil, South Africa, Indonesia, and Iran.

In its positioning for the Copenhagen meeting, China pledged that it would cut its carbon intensity 40-45% by 2020 from a 2005 baseline.[30] China's greenhouse gas intensity in 2005 ranked 34, indicating that its economy was comparatively GHG-inefficient, and suggesting that intensity reductions should be relatively easy to achieve. But in fact, based on CAIT data, this commitment simply reflects China's historical trend for intensity: for the 15-year period 1990 to 2005, China's carbon intensity, based on CO_2 emissions only, declined by 43% (its carbon- equivalent intensity, based on all six greenhouse gases, declined by 53%).[31]

Discussion

As stated above, the data on greenhouse gas emissions highlight issues of both effectiveness and fairness with respect to current efforts to address global climate change. Differentiating responsibilities between Annex I and non-Annex I countries fails to focus efforts on all the largest emitters. In addition, contradictory issues of fairness arise, as Annex I countries bear essentially all the direct economic costs of reducing emissions, and non-Annex I countries are granted the right to increase emissions to meet developmental needs. Finally, the focus on historical emissions as a baseline for regulation has differential and arbitrary impacts on individual nations.

The result of the UNFCCC and Kyoto Protocol's setting emissions targets for only developed nations and focusing on returning their emissions to a specific baseline is twofold: (1) the current regime has had little effect on global emissions, and will have little effect in the near future; and (2) the largest emitters, the United States and China, have not found it in their interests to join in the international effort to a significant degree. Indeed, the United States pulled completely out of the Kyoto process under the George W. Bush administration. This process has continued to be difficult, as the recent Copenhagen meetings illustrate.

This history of the UNFCCC and the Kyoto Protocol raises serious questions about how to develop greenhouse gas targets, time frames, and implementation strategies. With respect to targets, the UNFCCC recognized the right of developing countries to develop and the responsibility of all countries to protect the global climate. These goals of the UNFCCC suggest that if there is to be any permanent response to climate change that involves controlling greenhouse gases, then a regime that combines some measure reflecting the right of developing countries to develop, such as per capita emissions, and some measure reflecting the need to be efficient, such as carbon intensity, may be necessary to move the world toward a workable and effective climate change framework.

As shown above a global target focused on per capita emissions generally rewards developing nations,[32] providing them room for economic growth; the target's balance between limiting emissions and permitting growth determines the individual winners and losers. For example, based on **Appendix B**, a target of 3 tons carbon per person would allow all the developing nations in the top 20 emitters except South Korea growth room (South Korea is at 3.1 tons per capita), while five developed nations (United States, Russian Federation, Germany, Canada, and Australia) would have to make cuts. In contrast, a target focused on greenhouse gas intensity would have more diverse implications for developing nations. Several major developing nations produce considerably higher greenhouse gas emissions per million dollars of GDP than some developed nations. For example, in 2005 China's carbon intensity (372 tons/million international $GDP) was about four times that of the United Kingdom's (91) and Italy's (93). Thus a greenhouse gas intensity goal could be a counterforce to the economic development process for some countries, meaning that the winners and losers of a regime combining per capita and carbon intensity measures could be highly dynamic and contentious. Adding land-use implications would further complicate the regime, and selectively affect certain nations, especially those just now at the point of exploiting forests (notably Indonesia and Brazil).

For the United States, a regime containing some mix of per capita and greenhouse gas intensity measures[33] would likely imply a need to constrain emissions over some time frame. The U.S. greenhouse gas intensity is declining, as is the case with most nations, but the decrease currently does not completely offset increased emissions resulting from the growth of population and of the economy. The extent to which targets could translate into economic costs would depend on the other two features of the regulatory scheme: (1) time frame (specifically, whether it would accommodate technological advances in less-carbon-intensive technology or accelerated commercialization of existing low-carbon technologies such as nuclear power); (2) implementtation strategy (specifically, whether it encourages least-cost solutions and development of advanced technologies).

With respect to time frame, the data indicate two things: (1) most countries that achieved a significant reduction during the 1990s did so as a result of either an economic downturn or a substantial realignment in energy policy; (2) many countries have not been able to stabilize their emissions despite the UNFCCC's voluntary goal, much less reduce them. That failure was the impetus for the Kyoto Agreement's prescribed reductions and of the Copenhagen meeting. Using economic contraction as an emission reduction strategy can scarcely be considered an option. Instead, the substantial development and/or deployment of less-carbon-intensive technology, improved land-management strategies, and other actions would be necessary to achieve stabilized emissions. As noted above, greenhouse gas emissions are closely tied to industrialization—a synonym for "developed." With few exceptions, improvement in efficiency has been gradual. A permanent transformation of the global economy necessary to ensure a long-term stabilization of greenhouse gas emissions may involve a multi-stage, long-term time frame.

The difficulty in implementing the UNFCCC suggests implementation and compliance are still an open issue. The United States submitted climate action plans during the 1990s indicating it would achieve the UNFCCC goal of returning emissions to 1990 levels. It did not. There were no sanctions. Likewise, some Kyoto signatories may not achieve their reduction targets in 2008-2012. The sanctions are unclear. Now, for the Copenhagen Accord, nations are asked to voluntarily commit to reductions. Given the wide range of situations illustrated by the data, a flexible strategy that permits each country to play to its strengths may make it easier for diverse countries like the United States and China to reach some acceptable agreement.

The extent of flexibility would depend on the balance between emission reductions and economic cost designed into the targets, time frame, and implementation strategy. Market-based mechanisms to reduce emissions focus on specifying either the acceptable emissions level (quantity), or compliance costs (price), and allowing the marketplace to determine the economically efficient solution for the other variable. For example, a tradeable permit program sets the amount of emissions allowable under the program (i.e., the number of permits available caps allowable emissions), while permitting the marketplace to determine what each permit will be worth. Conversely, a carbon tax sets the maximum unit (per ton of CO_2) cost that one should pay for reducing emissions, while the marketplace determines how much actually gets reduced.

Hence, a major implementation question is whether one is more concerned about the possible economic cost of the program and therefore willing to accept some uncertainty about the amount of reduction received (i.e., carbon taxes), or one is more concerned about achieving a specific emission reduction level with costs handled efficiently, but not capped (i.e., tradeable permits). Of course, combinations of these approaches are possible, depending on the flexibility desired.[34] The data presented here portray a very wide range of situations and conditions among the 20 top countries that represent over 70% of total emissions. Significant flexibility may not only be desirable but necessary for them to reach any significant agreement.

End Notes

[1] For background, see CRS Report RL345 13, *Climate Change: Current Issues and Policy Tools*, by Jane A. Leggett.

[2] For the most recent developments on submissions to the UNFCCC by non-Annex 1 countries, see http://unfccc.int/national_reports/annex_i_ghg_inventories/inventory_review _reports/items/4715.php.

[3] The UNFCCC divides nations into two groups, nations listed in Annex I (which under the Kyoto Protocol would have specified reduction targets), encompassing "developed" nations including Eastern Europe and the former Soviet Union; and non-Annex I nations (which do not have specified reduction targets), including the rest of the world.

[4] Called the Climate Analysis Indicators Tool (CAIT), the database uses a variety of data sources to provide information on greenhouse gas emissions, sinks, and other relevant indicators. Full documentation, along with caveats, is provided on the WRI website at http://cait.wri.org/.

APPENDIX A. RELATIVE RANKING OF 20 TOP EMITTERS (PLUS EU-27) OF GREENHOUSE GASES BASED ON 2005 GREENHOUSE GAS EMISSIONS

Table A-1

Country	Annex I	2005 GHG Emissions (without land use)	1990 GHG Emissions (without land use)	2005 Per Capita GHG Emissions (without land use)	2005 GHG Intensity (without land use)	2005 GHG Emissions (with land use)	1850-2006 Cumulative Energy CO_2 Emissions (without land use)
China	No	1	2	81	34	1	2
United States	Yes	2	1	9	122	2	1
European Union-27[a]	*Yes*[b]	*[3]*[c]	*[2]*	*[43]*	*[154]*	*[3]*	*[2]*
Russian Federation	Yes	3	3	22	45	5	3
India[a]	No	4	6	148	88	6	8
Japan[a]	Yes	5	5	39	166	7	6
Brazil	No	6	9	84	107	3	21
Germany[a]	Yes	7	4	28	157	8	4
Canada	Yes	8	10	10	103	9	9
United Kingdom[a]	Yes	9	8	38	170	11	5
Mexico	No	10	13	72	134	10	15

Table A-1 (Continued)

Country	Annex I	2005 GHG Emissions (without land use)	1990 GHG Emissions (without land use)	2005 Per Capita GHG Emissions (without land use)	2005 GHG Intensity (without land use)	2005 GHG Emissions (with land use)	1850-2006 Cumulative Energy CO_2 Emissions (without land use)
Indonesia	No	11	17	117	77	4	25
Korea (South)[a]	No	15	19	29	128	12	20
Italy[a]	Yes	13	12	48	167	13	12
Australia[a]	Yes	16	15	6	71	14	14
Iran[a]	No	12	22	59	72	15	23
France[a]	Yes	14	11	51	179	16	7
Ukraine[a]	Yes	17	7	40	18	17	10
Spain[a]	Yes	18	20	44	159	20	17
South Africa[a]	No	19	16	50	53	22	13
Turkey	Yes	20	21	83	133	21	29

Source: Climate Analysis Indicators Tool (CAIT) Version 7.0. (Washington, DC: World Resources Institute, 2009).

a. Data from Land Use Change & Forestry not available.

b. European Union members, listed in Annex I, signed the Kyoto Protocol individually and, collectively, as the EU. The Protocol gave explicit authority to the original 15 member European Union to meet its obligations collectively; the EU has coordinated the compliance strategies of the newer member states into its overall compliance scheme, but those countries retain their individual Kyoto reduction targets.

c. The bracketed numbers would be the ranking of the EU; if the EU ranking were counted, equal and lower rankings would increase by one (e.g., Turkey would rank 21st in 2005 emissions and 84th in 2005 per capita emissions).

Appendix B. Greenhouse Gas Emissions and Other Climate Change-Related Indicators for 2005 Top 20 Emitting Countries (Excludes Land Use Change & Forestry)

Table B-1

2005 Rank	Country	Annex 1	2005 GHG Emissions MMTCE	2005 GHG missions % of World	1990 GHG Emissions MMTCE	1990-2005 Emissions Difference MMTCE	1990-2005 Increase or Decrease %	2005 Per Capita GHG Emissions (tons C/person)
1	China	No	1,974	19.1%	981	993	101.2%	1.5
2	United States	Yes	1,892	18.3%	1,634	258	15.8%	6.4
[3]	*European Union-27*	*Yes*[a]	*1,378*	*13.4%*	*1,467*	*-89*	*-6.1%*	*2.8*
3	Russian Federation	Yes	532	5.2%	800	-268	-33.5%	3.7
4	India	No	509	4.9%	302	207	68.5%	0.5
5	Japan	Yes	370	3.6%	326	44	13.5%	2.9
6	Brazil	No	276	2.7%	188	88	46.8%	1.5
7	Germany	Yes	266	2.6%	326	-60	- 18.4%	3.2
8	Canada	Yes	202	2.0%	159	43	27.0%	6.2
9	United Kingdom	Yes	176	1.7%	194	-18	-9.3%	2.9
10	Mexico	No	176	1.7%	125	51	40.8%	1.7

Table B-1 (Continued)

2005 Rank	Country	Annex 1	2005 GHG Emissions MMTCE	2005 GHG missions % of World	1990 GHG Emissions MMTCE	1990-2005 Emissions Difference MMTCE	1990-2005 Increase or Decrease %	2005 Per Capita GHG Emissions (tons C/person)
11	Indonesia	No	159	1.5%	90	69	76.7%	0.7
12	Korea (South)	No	155	1.5%	84	71	84.5%	3.2
13	Italy	Yes	154	1.5%	137	17	12.4%	2.6
14	Australia	Yes	153	1.5%	110	43	39.1%	7.5
15	Iran	No	152	1.5%	67	85	126.9%	2.2
16	France	Yes	150	1.4%	147	3	2.0%	2.5
17	Ukraine	Yes	135	1.3%	257	-122	-47.5%	2.9
18	Spain	Yes	119	1.2%	76	43	56.6%	2.7
19	South Africa	No	115	1.1%	91	24	26.4%	2.5
20	Turkey	Yes	107	1.0%	71	36	50.7%	1.5
Total[b]			7,772	75.3%	6,165	1,607	26.1%	
	WORLD		10,320	100.0%	8,380	2,189	26.1%	1.6`

Source: Climate Analysis Indicators Tool (CAIT) Version 7.0. (Washington, DC: World Resources Institute, 2009).

a. The Kyoto Agreement gave explicit authority to the original 15 member European Union to meet its obligations collectively; the EU has in effect expanded that authority as it has incorporated new members. If the EU-27 were ranked in terms of its 2005 GHG emissions, it would place 3rd.

b. Totals are of the 20 individual nations; they do not include the European Union.

Appendix C. Additional Emissions and Other Climate Change-Related Indicators for 2005 Top 20 Emitters

Table C-1

2000 Rank	Country	2005 GHG Emissions (without land use) (MMTCE)	2005 GHG Emissions (with land use) (MMTCE)	1950-2000 Cumulative Energy CO_2 Emissions (without land use) (MMTCE)	2005 Per Capita GDP (millions of international $)	2005 GDP (millions of international $)	2005 GHG Intensity (without land use) (tons/million intl. $GDP)
1	China	1,974	1,962	27,075	$4,076	$5,314,365	372
2	United States	1,892	1,860	91,088	$41,873	$12,376,100	153
[3]	*European Union-27*[ab]	1,378	*1,378*	*83,447*	*$26,883*	*$13,175,925*	*105*
3	Russian Federation	532	547	25,404	$1 1,861	$1,697,957	313
4	India[a]	509	509	7,487	$2,234	$2,445,194	208
5	Japan[a]	370	370	12,155	$30,310	$3,872,843	96
6	Brazil	276	776	2,581	$8,505	$1,582,642	174
7	Germany[a]	266	266	21,937	$31,397	$2,589,299	103
8	Canad[a]	202	219	6,860	$35,065	$1,133,018	178
9	United Kingdom[a]	176	176	18,623	$32,207	$1,939,686	91
10	Mexico	176	186	3,212	$12,563	$1,295,157	136
11	Indonesi[a]	159	557	1,788	$3,197	$705,159	226
12	Korea (South)[a]	155	155	2,699	$22,783	$1,096,741	142

Table C-1 (Continued)

2000 Rank	Country	2005 GHG Emissions (without land use) (MMTCE)	2005 GHG Emissions (with land use) (MMTCE)	1950-2000 Cumulative Energy CO_2 Emissions (without land use) (MMTCE)	2005 Per Capita GDP (millions of international $)	2005 GDP (millions of international $)	2005 GHG Intensity (without land use) (tons/million intl. $GDP)
13	Italy[a]	154	154	5,132	$28,122	$1,648,164	93
14	Australia[a]	153	153	3,470	$31,702	$646,550	236
15	Iran[a]	152	152	2,211	$9,314	$643,503	236
16	France[a]	150	150	8,810	$30,710	$1,869,387	80
17	Ukraine[a]	135	137	6,856	$5,583	$263,007	512
18	Spain[a]	119	119	2,915	$27,366	$1,187,655	100
19	South Africa[a]	115	115	3,492	$8,504	$398,757	289
20	Turkey	107	116	1,490	$10,977	$781,243	136
Total[c]		7,772	7810	255,285		$43,486,427	
	WORLD	10,320	11,868	314,056	$8,769	$56,601,925	182

Source: Climate Analysis Indicators Tool (CAIT) Version 7.0. (Washington, DC: World Resources Institute, 2009).

Note: Due to rounding, independent calculations may give slightly different results.

a. Data from Land Use Change and Forestry not available.

b. The Kyoto Agreement gave explicit authority to the original 15 member European Union to meet its obligations collectively; the EU has coordinated the compliance strategies of the newer member states into its overall compliance scheme, but those countries retain their individual Kyoto reduction targets. If the EU-27 were ranked in terms of its 2005 GHG emissions, it would place 3rd.

c. Totals are of the 20 individual nations; they do not include the European Union.

[5] Both the individual countries of the European Union and the European Community as an entity are Parties to the Kyoto Protocol. Within the EU, the differing situations of each constituent nation have resulted in differing emissions targets and policies for each country. While this analysis focuses on the implications of individual nations' situations, fifteen member states of the EU are authorized to meet their goals collectively.

[6] Carbon intensity is the ratio of a country's emissions to its gross domestic product (GDP), measured in international dollars (purchasing power parity).

[7] Carbon dioxide, nitrous oxide, methane, perfluorocarbons, hydrofluorocarbons, and sulfur hexafluoride.

[8] *United Nations Framework Convention on Climate Change*, Article 3.1.

[9] The United States and many other countries failed to meet this voluntary goal. It was this general failure that gave impetus to the Kyoto Protocol to mandate reductions.

[10] Generally the baseline was 1990; the individual Annex I commitments were negotiated, with the U. S. commitment—*if* the United States had agreed to the Kyoto Protocol—being a 7% reduction.

[11] ClimateAnalysis Indicators Tool (CAIT) Version 7.0 (Washington, DC: World Resources Institute, 2009).

[12] http://en.cop15.dk/files/pdf/copenhagen_accord.pdf

[13] For a more general discussion of the top 25 emitters in the year 2000, see Kevin Baumert and Jonathan Pershing, *Climate Data: Insights and Observations* (Pew Center on Climate Change, December 2004).

[14] CAIT's EU-27 includes the EU-15 (Austria, Belgium, Denmark, Finland, France, Germany, Greece, Ireland, Italy, Luxembourg, Netherlands, Portugal, Spain, Sweden, United Kingdom), plus Bulgaria, Cyprus, Czech Republic, Estonia, Hungary, Latvia, Lithuania, Malta, Poland, Romania, Slovakia, Slovenia.

[15] The UNFCCC provides a methodology for calculating the greenhouse gas contributions of nations and converting them to equivalent units—Million Metric Tons of Carbon Equivalents (MMTCE).

[16] However, for CO_2 only, the United States remained the leading emitter in 2005.

[17] Of the top 20 in1990, 18 are still in the top 20 15 years later, albeit with some shifting in order (most notably, China edging ahead of the United States in total greenhouse gas emissions). Kazakhstan dropped out of the top 20 early in the 1990s, and was replaced by Iran. Between 2000 and 2005, the only change in the top 20 was Turkey slipping ahead of Poland for the 20th spot.

[18] For a discussion of these situations, see CRS Report RL33970, *Greenhouse Gas Emission Drivers: Population, Economic Development and Growth, and Energy Use*, by John Blodgett and Larry Parker.

[19] Germany falls into this category as a result of its incorporation of East Germany. The pre-merger West Germany was of course not a centrally planned economy.

[20] Kazakhstan and Poland, which were in the top 20 in 1990, also fall into the Annex I nations with declining emissions; with the decline of their coal based economies, they dropped out of the top 20, ranking 32 and 22 respectively, in 2005. Together they accounted for about 1.5% of 2005 emissions.

[21] France's emissions declined between 1990 and 2000, and between 1990 and 2005 increased only 2.0%.

[22] Note that the cumulative data includes only energy-related CO_2.

[23] See CRS Report RS22964, *Measuring and Monitoring Carbon in the Agricultural and Forestry Sectors*, by Ross W. Gorte and Renée Johnson.

[24] E.g., the availability of natural gas and/or coal, and when each has been or is being exploited; or the extent of deforestation and/or afforestation, and when either has occurred.

[25] For other analyses bearing on this question, see CRS Report RL32762, *Greenhouse Gases and Economic Development: An Empirical Approach to Defining Goals*, by John Blodgett and

Larry Parker; and CRS Report RL33970, *Greenhouse Gas Emission Drivers: Population, Economic Development and Growth, and Energy Use*, by John Blodgett and Larry Parker.

[26] The top five by this measure are countries whose economies are dominated by oil and gas production.

[27] While the term "greenhouse gas intensity" encompasses all six greenhouse gases, the term "carbon intensity" is sometimes used identically and implicitly means "carbon equivalents intensity" and other times is used more narrowly to refer only to carbon emissions. The discussion in this analysis focuses on "greenhouse gas intensity," unless otherwise noted (e.g., in the discussion of cumulative emissions).

[28] See Kevin A. Baumert, Ruchi Bhandari, and Nancy Kete, *What Might A Developing Country Climate Commitment Look Like?* World Resources Institute Climate Notes, May 1999.

[29] See CRS Report RL33970, *Greenhouse Gas Emission Drivers: Population, Economic Development and Growth, and Energy Use*, by John Blodgett and Larry Parker.

[30] http://en.rian.ru/world/20091126/156994803.html; http://www.guardian. http://www.guardian..co.uk/environment/2009/nov/26/china-targets-cut-carbon-footprint

[31] China's commitment on intensity has parallels with George W. Bush's 2002 call for reducing intensity 18% by 2012 – a reduction that was only slightly more aggressive than the business-as-usual trend.

[32] An exception is several oil-producing States, mainly in the Gulf, that are high emitters due to exploitation of their oil reserves.

[33] See CRS Report RL33970, *Greenhouse Gas Emission Drivers: Population, Economic Development and Growth, and Energy Use*, by John Blodgett and Larry Parker.

[34] See CRS Report RL33799, *Climate Change: Design Approaches for a Greenhouse Gas Reduction Program*, by Larry Parker; CRS Report RL30024, *U.S. Global Climate Change Policy: Evolving Views on Cost, Competitiveness, and Comprehensiveness*, by Larry Parker and John Blodgett; and CRS Report RS21067, *Global Climate Change: Controlling CO_2 Emissions—Cost-Limiting Safety Valves*, by Larry Parker.

INDEX

S

T

U

V

W